D0765652

Trying Biology

Trying Biology

The Scopes Trial, Textbooks, and the Antievolution Movement in American Schools

ADAM R. SHAPIRO

The University of Chicago Press

CHICAGO AND LONDON

ADAM R. SHAPIRO is a lecturer in intellectual and cultural history at Birkbeck, University of London.

The University of Chicago Press, Chicago 60637
The University of Chicago Press, Ltd., London

Printed in the United States of America

22 21 20 19 18 17 16 15 14 13 1 2 3 4 5

ISBN-13: 978-0-226-02945-0 (cloth)
ISBN-13: 978-0-226-02959-7 (e-book)

Library of Congress Cataloging-in-Publication Data
Shapiro, Adam R.
Trying biology : the Scopes trial, textbooks, and the antievolution movement in American schools / Adam R. Shapiro
pages ; cm
Includes bibliographical references and index.
ISBN 978-0-226-02945-0 (Cloth : alk. paper)—
ISBN (invalid) 978-0-226-02959-7 (e-book)
1. Evolution (Biology)—Study and teaching—United States—History. 2. Scopes, John Thomas—Trials, litigation, etc. 3. Biology—United States—Textbooks—History. 4. Biology publishing—United States—History. 5. Religion and science—United States—History. I. Title
QH362.S53 2013
576.8—DC23
2012042417

♾ This paper meets the requirements of ANSI/NISO Z39.48-1992 (Permanence of Paper).

Contents

CHAPTER ONE • 1
Beyond Science and Religion:
The Scopes Trial in Historical Context

CHAPTER TWO • 14
The Textbook Trust and State Adoption

CHAPTER THREE • 39
Textbooks and Their Makers:
Authors, Editors, Salesmen, and Readers

CHAPTER FOUR • 62
Civic Biology and the Origin of the
Antievolution Movement

CHAPTER FIVE • 87
How Scopes Was Framed

CHAPTER SIX • 111
The Evolution of the *New Civic Biology*

CHAPTER SEVEN • 135
Biology Textbooks in an Era of
Science and Religion

CHAPTER EIGHT • 157
Losing the Word:
Measuring the Impact of Scopes

Acknowledgments • 167
Notes • 169
Index • 189

CHAPTER ONE

Beyond Science and Religion: The Scopes Trial in Historical Context

It was late 1923 when Tennessee governor Austin Peay took note of a deadline that would shape the future of his state and help define his political career. On September 1, 1924, textbook contracts would expire. The state had regulated the adoption of textbooks since 1899, standardizing the titles and prices of school texts in five-year cycles. The new adoption itself was nothing unprecedented. But people in Tennessee would experience an astounding increase in textbook prices, possibly even as much as 50 percent.

The cost of new textbooks had actually increased in 1920 and 1921 as a result of post–World War I inflation and commodity shortages. The costs of paper, binder's board, and cloth had more than doubled, and labor costs had increased nearly as much.[1] But Tennessee's citizens were lucky: the state had locked in prices with contracts signed in 1919. Many neighboring states were also in the practice of adopting textbooks in five-year cycles but had entered new contracts in 1922 or 1923. People in those states were already paying more for the same titles. In September, Tennessee's advantage would come to an end.[2]

Tennessee's 1919 adoption came at a fortuitous time for its citizens and schools but expired at an awkward time for Austin Peay and his drive for school reform in the state. Just two months after Tennesseans would feel the impact of the price increase, Peay would stand for reelection. He had already presented himself as an advocate of public education, and he planned to make the expansion and improvement of the state's schools the defining issue of his 1924 campaign. Tennessee ranked among the worst of the states in literacy and in economic production. Much of Peay's two-year first term had been dedicated to modernizing the state's infrastructure, enabling it to industrialize and become more integrated within the national economy. The next phase of his plans for state development was tied to education and to using the schools to train a new generation to work in this new environment.

A drastic increase in textbook prices would have crippled political support for the additional taxes necessary for school reform. Except in a few large cities where textbooks were purchased by school districts, students (or their parents) usually bought textbooks from designated local depositories at contractually fixed prices. While the greater expense probably would not have cost Peay reelection, it could have reduced his support in a legislature that had already forced the governor to scale back some of his ambitions for the state.

But with the dawning of the new year came an inventive solution to this problem. Perry L. Harned, the commissioner of education and a political ally from Peay's hometown of Clarksville, helped devise a strategy to reduce the impact of this price increase. On January 3, 1924, Peay wrote Harned:

> We must not consider an advance of 33-1/3% in our school books. I will exhaust every means available before consenting to such a contract.
>
> I request you to communicate immediately with the publishers having contracts now with the state and express to them my view that we are entitled, for the school term 1924–25, to buy our books under existing contracts since the schools will be running before the contracts will expire.[3]

Many schools in the state would be opened in August, the rest by early September. Any textbooks bought before September 1 would be available for the prices set in the 1919 contract. Keeping the same books also meant that books could be bought secondhand or used again by families with multiple children. If people bought textbooks promptly, a new adoption would not be needed for another year. In effect, Tennessee could squeeze a sixth year out of a five-year contract.

Peay and Harned did not announce this plan right away, and, as the adoption's expiration drew nearer, others took note of the looming rise in textbook prices. In February, a legislator from a rural district solicited Peay's "influence in not changing school books this year, as it will be a heavy expense on the people, especially farmers."[4] In March, Peay privately confided to some acquaintances that he was "determined to have no adoption until next year."[5] But there was no public announcement until April 16, 1924, when the state textbook commission, appointed by the governor, "unanimously decided to defer the adoption of text books until 1925."[6] By delaying their announcement for several months, Peay and Harned effectively (though perhaps not intentionally) allowed the perception of a crisis to build before unveiling their solution.

Saving Tennesseans from higher textbook prices, they were hailed as champions of poor farmers and urban school reformers alike. The trouble Peay had identified in December had become a political success story by April. Voters would not feel the effects of a price increase until after the election. The strategy of coaxing an extra year out of the 1919 contract cast Peay as a governor who saved taxpayers money while promoting and defending education. From September 1, 1924, through June 30, 1925, the books adopted in 1919 would be used. If anyone needed to buy new textbooks, they would be available at increased prices, but most people either bought new books early or were able to buy or reuse old copies. Shortly after the plan was implemented, Harned wrote: "Practically all the books needed for the year were bought before September 1st and . . . the children here and there needing books since Sept. 1 bought second hand books in the community. Book men report practically no business since September 1st."[7] Nearly 90 percent of students in Tennessee began the school year before September 1.[8] Tennesseans saved hundreds of thousands of dollars during the 1924–25 school year because the new adoption was postponed. In a published list of interim prices for that school year, Harned claimed that, in addition to this year's savings, it would be beneficial for future adoptions to occur around July 1, at the beginning of the scholastic year, instead of September 1, when many students had already begun classes.[9] He presented the postponed adoption as part of his intended reforms of the administration of Tennessee's public schools.

Postponing the 1924 textbook adoption in Tennessee saved school parents a huge expense and bolstered support for state expansion of education. Peay won reelection in 1924 with a mandate to pursue this agenda. For everyone (except perhaps textbook publishers) the postponement was a good thing. It did, however, contribute directly to one unforeseen consequence: the Scopes trial.

In the waning days of the 1924–25 school year, John Scopes, filling in for the regular biology teacher at Rhea County Central High School in Dayton, Tennessee, assigned reading from George W. Hunter's 1914 textbook *A Civic Biology* to prepare students for their final examination.[10] The consequences of his doing so have become legendary. Scopes volunteered to be tried in a test case meant to challenge the validity of a new antievolution law enacted by Tennessee in March 1925. He was indicted and convicted and became a symbol of the (perceived) ongoing warfare between science and religion. He was hailed by some as a martyr, compared to the likes of Socrates and Galileo. Often linked to the epithet Monkey Trial, the name Scopes also became a label of derision employed by those who saw evolution as an irreligious and immoral doctrine.

The Scopes trial is one of the best-known events in the history of science and religion in the United States. Subsequent trials concerning the teaching of evolution have been referred to as "Scopes II," and even court cases in other countries have been compared to Scopes. The trial's fame stems in part from the participation of two men—the acclaimed attorney Clarence Darrow and the prominent politician William Jennings Bryan. Their debate over the truth of the Bible and its relation to science captivated the country and was reported throughout the world. Because of this spectacle, this misdemeanor trial in a rural county seat in Tennessee has often been referred to as one of "the trials of the [twentieth] century."[11] In the twenty-first century, interest in the trial has shown no sign of waning. Evolution, and its presence in schools, has become even more politically polarized and more deeply connected to claims of conflict between science and religion than it was in the 1920s. The Scopes trial introduced new ways of thinking about science and religion and their roles in public life throughout the United States and even across the globe.[12]

The assumption that the Scopes trial was rooted in the fundamental incompatibility of science and religion—and was therefore inevitable—has ensured its continued invocation in popular culture. This conflict narrative was cultivated by the trial's participants and has been invoked by those who co-opted the trial to justify later conflicts over evolution.[13] This has led to a historical cycle of self-justification. Science-religion conflict explains why there was a Scopes trial, and the Scopes trial proves the reality of science-religion conflict. To accept this is to embrace the explanation put forth by the trial's participants, who of course accounted for the trial in ways that justified and valorized their own involvement. This self-justification deserves to be viewed with skepticism, and the idea that the Scopes trial *proves* that science and religion are at loggerheads must be called into question.

This book is an attempt to break the cycle of self-justification that permeates historical accounts of science and religion. It asks how we can give an account of an apparent conflict—the Scopes trial—without resorting to a claim that evolution and the Bible are simply incompatible. To explain how this social movement against the teaching of evolution culminated in a spectacle of science-religion conflict, this book explores how a new high school curriculum—called *biology*—was developed, printed, distributed, and sold by authors, publishers, educators, and government officials; what that curriculum contained; and how it became an instrument of political ideology tied to an expanded role for compulsory schooling in shaping American life. It then investigates how and why the conflict over American education became understood as a battle over science and religion by examining the ways in which participants in the Scopes trial and producers and consumers of science textbooks reinterpreted the debate over evolution.

DEBUNKING THE INEVITABLE SCOPES TRIAL

Given the trial's importance to the history of science and religion in America, it may seem surprising that if it were not for the 1924 postponement of textbook adoptions in Tennessee—or the postwar jump in commodity prices that prompted it—there likely would not have been a Scopes trial. But it is clear that postponing the adoption until 1925 left an eleven-year-old biology textbook in the hands of ninth and tenth graders. When it was first published, *A Civic Biology* was a groundbreaking textbook, and for many years it was a best seller. Several thousand copies were sold in Tennessee after its adoption in 1919, but by the early 1920s its national sales had started to wane as newer textbooks competed with it.[14] It continued to sell because of the long term of some adoptions, but by 1924 it had been on the market for ten years; its publisher, the American Book Company (ABC), already felt it overdue for a revision. For reasons of its age alone, the *Civic Biology* would not have been readopted in Tennessee—had an adoption occurred in 1924. In neighboring Kentucky, where an adoption did take place in the summer of 1924, Hunter's *other* textbook, the much more recent *New Essentials of Biology* (1923), was selected.[15]

It is speculative to suggest what *might* have happened had history unfolded differently, but to suggest that the political circumstances that left Hunter's *Civic Biology* in Tennessee high schools in 1925 were not critical to the Scopes trial is to accept that something like the trial was bound to happen regardless—that it was, in a sense, inevitable. This myth of inevitability has long been part of the story of the Scopes trial. Many people believe that the conflict of science and religion was so intractable, especially over the question of human origins, that from the moment Charles Darwin penned *The Origin of Species* something like the Scopes trial was all but certain. According to this interpretation of the trial, the specific details of the case were just happenstance. The *essential* conflict between evolution and Christianity was unavoidable. If the science and religion conflict had not come to a head in the legal case of *Tennessee v. John Scopes* in 1925, something similar would have occurred in another setting. As a result, there was nothing *unusual* about Dayton, Tennessee, or John Scopes that brought about the Scopes trial; the town and the man were subject to historical forces much larger than they were.

The truth is that this myth of inevitability was a self-serving invention of the Scopes trial's participants. Antievolution sentiment was growing in America, but that movement was finding other, less confrontational outlets. Some high school textbooks published in the early 1920s—books that Tennessee might have considered in 1924 and that had been adopted in many other states—approached the topic of evolution differently or avoided discussion of it. Most of these books were published after William Jennings Bryan launched his attack on the teaching of evolution in 1922. One publisher even consulted Bryan on how to present the subject in an inoffensive way.[16] If not for *Tennessee v. Scopes,* antievolutionism might not have had another chance to culminate in a large public spectacle.

Perhaps even more devastating to the myth of inevitability is the fact that the conditions for such a public spectacle had all been met before, without anything like the Scopes trial resulting. There was another antievolution trial that took place *a year before* Scopes. David S. Domer, a teacher in Nebraska, was fired from a Lutheran college after church members in the town where he had taught complained that he was "unfit morally and mentally" because he was an evolutionist. In 1924, he sued these church members for slander and was awarded over $5,000 in damages.[17] If teaching evolution was the main source of outrage that was needed to create a public battle between religion and science, then why did it not happen in Nebraska? In some ways, the Domer case would have been a more natural venue to debate the morality of teaching evolution. The trial even took place in Lincoln, where William Jennings Bryan had published a newspaper and which he had represented in Congress. Perhaps the Tennessee case should have been known as "Domer II." Instead, that trial was ignored outside Nebraska at the time and remains unknown today. Its obscurity demonstrates that *Tennessee v. Scopes* was more than a mere expression of an unavoidable science and religion clash. There were unique circumstances in Tennessee that made a public event possible, circumstances that may not have concatenated anywhere else.

The continued use of Hunter's obsolescent *Civic Biology* was a crucial part of those circumstances. Scopes recounted in his memoir: "I didn't know, technically, whether I had violated the law or not." He was reassured by consulting the *Civic Biology*. "There's our text, provided by the state. I don't see how a teacher can teach biology without teaching evolution."[18] Just a month before the Scopes trial, William Jennings Bryan wrote to a colleague: "If you have not read the book in question, 'Hunter's Civic Biology', I suggest you get it. It certainly gives us all the ammunition we need."[19] There were few other states where such aging materiel could still be found in 1925.

Yet, in signing the antievolution law, Governor Peay wrote: "I can find nothing of consequence in the books now being taught in our schools with which this bill will interfere in the slightest manner."[20] Though Peay and Bryan saw eye to eye on many political issues, they obviously did not *interpret* the *Civic Biology* in the same way. It was not merely the presence of Hunter's book that brought about the trial but a clash between groups of readers who advocated different interpretations of what the textbook said and what was meant by teaching "evolution."

There was no sudden realization in 1925 that Darwin's theories, over half a century old, conflicted with essential truths of religion. Yet most histories of the trial begin from the assumption that the trial was fundamentally about the relationship of science and religion.[21] This treatment suggests an event that is freed from the specifics of time and place. As an epic science-religion clash, the trial, or something very similar, could occur almost anywhere, at any time. It is that view that allows latter-day commentators to refer to the next "Scopes trial" whenever any antievolution case makes the news. That the *actual* Scopes trial was in Dayton is almost accidental to the matters at hand. The trial was more than an expression of some eternal struggle; it was a specific event, and its occasion relied on several different factors. Some of those contributing factors were unique to Tennessee and even to Dayton, and some had little or nothing to do with either science or religion. The postponement of the 1924 textbook adoption may at first seem a triviality in the series of events that led to the Scopes trial, but, once the role of political maneuvering and the regulation of education in Tennessee are better understood, the event proves crucial.

In order to understand how the Scopes trial became a signature event in the history of science and religion, it is necessary to look beyond science and religion for an explanation. To understand why Hunter's book was in John Scopes's hands, it is necessary to examine the history of textbook adoption and publishing. In order to understand the motivations behind the antievolution law, it is necessary to understand the history of biology curricula and school reform policy. It is in these contexts that the not-so-inevitable Scopes trial came to be.

WHY DAYTON OF ALL PLACES?

When Tennessee adopted Hunter's *Civic Biology* in 1919, it was set for schools throughout the state. While its focus on the use of biology in urban environments might have been well suited to Memphis, Nashville, or Chattanooga, its use in Dayton was less appropriate. In the 1890s and the first decade of the twentieth century, Dayton had been a booming center of the coal-oil industry, with a population that had swelled to over four thousand; by 1925, most of the industry had dried up, and Dayton's population had shrunk to eighteen hundred.[22] Tension between urban and rural interests was not unusual in state politics, and in Tennessee rural Rhea County (where Dayton was located) was compelled to use biology textbooks that touted the benefits of urban life and prepared students to be city dwellers. Rhea County was not unique in using this textbook, but, because of its earlier growth and affluence, it was one of the few rural counties that had public high schools prior to 1925. Rhea County Central High School had been built in Dayton in 1906, during the peak of the industrial boom.[23]

Before the trial was under way, people were already asking, Why Dayton of all places? Some of the town's leading citizens even issued a pamphlet asking the very question (see fig. 1). In the pamphlet, they tried to portray Dayton as *the* typical American town, quoting from Sinclair Lewis's *Main Street*: "This is America—a town of a few thousand, in a region of fruit and corn and dairies and little groves."[24] Pithily, the pamphlet turned the question around: "'Why Dayton—of *all* places?' You ask! And Dayton answers: 'Of all places, *why not Dayton*?'"[25]

Such rhetoric fit the scheme of representing the trial as an epic clash. By claiming that Dayton was, in a sense, every American town, the pamphlet's authors made it clear that the Scopes trial was not an expression of Dayton's peculiarity; instead, the location of Dayton made the trial universally American. The debate between science and religion, between tradition and modernism, was the *American* debate. As Edward J. Larson and other Scopes trial historians have noted, the citizens of Dayton seized on the opportunity to host a test of the antievolution law as a means to publicize their town.[26] This portrayal of the trial as science-religion clash also served the need of those boosters who had an eye toward Dayton's economic revival. Dayton's leaders were understandably concerned that the town might be portrayed as provincial and backward; claims that it was archetypically American could serve to deflect such caricatures.

The claim that Dayton was, in a sense, Main Street, USA, was, on one hand, an attempt to assert the town's character as quintessentially American.

FIGURE 1 The first copies of *Why Dayton of All Places?* on sale. Courtesy of Tennessee State Library and Archives.

On the other, it was also an attempt to frame the character of America as typically Daytonian. As *Why Dayton of All Places?* describes:

> Through the Appalachian trails leading out of the Carolinas and the Virginias migrated the first white settlers into the land of the Tennessee. Upon the Cherokees' hunting grounds that rambled over river bottom lands, valleys, ridges and mountains came the purest strains of sturdy Anglo Saxons, planting their standards. And their children developed the commonwealth where Dayton grew.
>
> They were all a rugged lot and their convictions were adamant. In these respects they were no different than the pioneers of America everywhere.[27]

The picture of the pioneers as rugged settlers of a wilderness, as pure Anglo-Saxon stock tied to the land and to farming, presented one view of American identity that was being challenged by changing demographics in late nineteenth- and early twentieth-century America. It was also an image that ignored the racial and economic diversity of the South since its settlement.

Even those who eagerly portrayed Dayton as an antimodern backwater, such as the *Baltimore Sun* journalist H. L. Mencken, were just as happy as

Dayton's leaders to see nothing peculiar in the town. "Dayton may be typical Tennessee, but it is not all of Tennessee,"[28] Mencken wrote, while extolling a more "civilized" contrast in nearby Chattanooga. If not quintessentially American, it was at the very least the epitome of rural America, of white communities in the southern United States, and representative of the nascent culture clash that the Scopes trial would come to embody. W. E. B. DuBois also echoed this sentiment, proclaiming: "Dayton, Tennessee is America: a great ignorant, simple-minded land, curiously compounded of brutality, bigotry, religious faith, and demagoguery."[29] For DuBois, the great ignorance of antievolutionists at Dayton was the same ignorance that was used to justify racism and the continuing persecution of African Americans in the southern United States.[30] Like its portrayal as Main Street, the images of Dayton as a stereotypical rural town, or a stereotypical southern town, or a stereotypical white town were gross caricatures. Yet for all parties—supporters and detractors of the town and of the antievolution law—there was no more useful answer than: "Why not Dayton?" Any other response would have diminished the importance of the trial by highlighting its location, rather than the eternal debate between science and religion. Yet it does matter that the Scopes trial happened in the specific town of Dayton, and perhaps it could have happened *only* because of circumstances in the town of Dayton.

There were factors specific to Tennessee—and to Dayton within Tennessee—that explain why this town became the setting of one of the most important events in American religious history. However, Dayton's religious character was not itself unique. Its citizens were members of the same denominations that predominated in the state. It was not the home of any seminaries, organizations, or especially charismatic religious leaders. Indeed, there is nothing in the *religious* history of Dayton up to 1925 that would have recommended it to one's attention. Nor was the town a particularly scientific location. There were no laboratories in Dayton, no universities. The unique circumstances that led to the Scopes trial *in Dayton* have virtually nothing directly to do with the town as a place of either science or religion. Creating the image of Dayton as the all-American town reinforced the perception that the trial was rooted in generic American themes that could be found at any place and at any time. The images of Dayton and of the Scopes trial reinforced one another as epic and essential.

Because Dayton was presented as a microcosm of the United States, the antievolution trial also came to be seen as an expression of the nationwide tensions between urban industrial and rural agricultural ideologies. Dayton was a town that combined elements of both worldviews. The trial's jury consisted almost entirely of farmers. George Rappleyea, the man who first conceived the idea of a test case in Dayton, was a mining engineer.[31] For all its pretense of Main Street, Dayton *was* unusual. But perhaps what made it relatively unique was the extent to which it contained elements of such different American ways of life. It was a town that had experienced the boom and bust of industrialization and that had begun to reinvent itself as a rural town just as the state began construction of highways and railroads, embracing policies designed for modernization. While few towns in the United States could boast all these elements, the contrasts of cultures made this particular town, if not a typical American town, a fitting example of a divided and changing American culture. In this respect, the trial in Dayton did represent universal themes.

These themes were also expressed quite specifically in the high school biology textbooks, as the whole notion of "civic" biology bespoke the increasing urbanization and industrialization of the United States. The debate over biology textbooks, which were sold nationally and adopted at the state level, brought the ideas of civic biology to places where such cultural change aroused suspicion. This presentation of the Scopes trial as a dramatic singular event obscures the more gradual history of biology education and the real successes of the antievolution movement that had already taken place. The Scopes trial was at most one important episode in a drama that was years in the unfolding. It marked neither the beginning nor the end of the antievolution movement. To understand how events unfolded and what their consequences were, one must look both to the peculiarities of Dayton and to events and people that never directly came to the Tennessee hills.

SCIENCE, RELIGION, AND THE USES OF THE SCOPES TRIAL

As a legal event, *Tennessee v. Scopes* is a historical oddity. The defense had no interest in claiming that John Scopes was innocent. This was intentionally planned so that the antievolution law itself could be tested in appeal (although the Tennessee Supreme Court eventually threw out Scopes's conviction on a technicality unrelated to the validity of the antievolution law).[32] With the verdict presumed, the trial readily shifted away from the specific actions of John Scopes to debates over the truth of evolution and its incompatibility with religion.

Though it has no legacy as a legal precedent, the Scopes trial has continued to echo in public consciousness. It has sounded in courtrooms around the United States as well. Unsurprisingly, subsequent legal debates over evolution and religion have also focused on education. One of the most recent examples of this came after a school board in Pennsylvania attempted to introduce "intelligent design" as an alternative to evolution in its high school biology classroom. In the opinion issued on December 20, 2005, in the case *Kitzmiller et al. v. Dover Area School District*, Judge John E. Jones III referred to a history of "religiously motivated groups [that] pushed state legislatures to adopt laws prohibiting public schools from teaching evolution." Citing the 1968 Supreme Court ruling against antievolution laws (like the one passed by Tennessee in 1925), Jones attributed this to an "upsurge of fundamentalist religious fervor of the twenties." A wider view of the Scopes trial does not contradict what Judge Jones referred to as the "history of religious opposition to evolution,"[33] but it does illustrate that that opposition, which culminated in the Scopes trial, was not *merely* the result of religious fundamentalism and that opponents of evolution were not necessarily driven by their religious views. The trial had origins in debates over American education that had little to do with either science or religion.

Its roots in education had little to do with the way in which the Scopes trial has been *used*. The assumption that it was fundamentally rooted in the incommensurability of science with religion allowed the trial's participants to shift conversation away from local school politics and gave later culture warriors grounds for claiming a historical precedent.[34]

There was religious opposition to "Darwinism" that claimed that it was inaccurate and that belief in it threatened morality. Such responses had been around since the publication of the *Origin of Species* (if not earlier). But religious antievolutionism had little to do with schools until the 1920s. School antievolutionism was part of a larger backlash against the expansion of compulsory public education, the use of science teaching to promote certain views of citizenship, and the role of textbook publishers and teaching organizations in standardizing curricula.

John Scopes was right to say that one cannot teach biology without teaching evolution, but not just in the sense that he meant. Opposition to evolution was part of this larger conflict over schools, but not merely as a component of the biology curriculum. Evolution was not just a topic in the life sciences; it was part of an ideology of social progress that led to the creation of books like Hunter's *Civic Biology* and to state policies that promoted public schooling. Antievolutionism as a school movement was part of a larger response to this ideology, and the rhetoric of science and religion conflict was used and adapted for the battle that came.

Much of the recent scholarship on the history of science and religion has criticized the simplemindedness of conflict narratives and has emphasized the complexity of relationships between scientific and religious ideas.[35] But this attempt at nuance has had to compete with the common perception of the Scopes trial. By this view, the fact that the nondescript town of Dayton vaulted into the spotlight in 1925 seems to show that *only* a genuine science-religion conflict could have brought a Scopes trial about. After the trial, the rhetoric of inherent conflict became widespread, giving the event a sense of gravity. The fate of John Scopes was important—not just to the eighteen hundred people of Dayton, but also to the world—because the locality of Dayton became universalized and the legal proceedings in its courthouse became the spectacle immortalized as the "Scopes trial."

CHAPTER TWO

The Textbook Trust and State Adoption

In an era when hotels were increasingly becoming important sites of cultural and political activity, the Edwards Hotel was perhaps the most important building in Jackson, Mississippi. The twelve-story Beaux Arts hotel, close to the train station, was the home of many lawmakers during the state legislative session. In the years following its opening in 1923, it was probably the site of more dealmaking than the state capitol a mile away.[1]

But it was another kind of power that made the building unique. It was one of the first in Jackson built with electrical outlets in the walls, and it was this feature that led Joseph Littlejohn to lay down on the floor of room 953 shortly after 5:00 P.M. on March 15, 1928. The thirty-four-year-old man placed a pillow under his chest. By his head was a tablet and a pencil, and by his left ear the electrical socket had been dismantled. A few minutes later, ear cupped over the hole where the outlet had been, Littlejohn began to write.

On the other side of the wall, two men in room 951 spoke in hushed tones. Littlejohn transcribed their conversation as one of them, W. F. Nihart, picked up the telephone and placed a long-distance call to Cincinnati, Ohio.

Littlejohn heard a voice exclaim: "Oh! don't be afraid of that. We are doing all that is humanly possible here." The call ended, and Nihart addressed his colleague, Henry L. Louis: "Dr. Howe is afraid if we leave here that Bilbo will catch us napping, and that while we have him whipped, to keep him that way."

Over the course of several weeks, Joseph Littlejohn and John D. Kleyle, New Orleans–based investigators for the Pinkerton Detective Agency, kept track of the conversations that took place among Louis, Nihart, two other coworkers and the visitors they brought. During that time, some of the most influential people in the state—legislators, newspaper editors, and business leaders—paid a visit to room 951 or to other hotel rooms Louis had rented.[2]

This scene may seem like the stuff of spy novels, and an audience might expect the stakes in this intrigue to be appropriately world changing. In truth, this episode was cast from a more mundane mold: Louis and Nihart were textbook salesmen for the American Book Company (ABC). They had come to Mississippi to influence a bill before the state legislature. They were sent by W. T. H. (William Thomas Hildrup) Howe, the ABC's managing partner in charge of sales in the South. Howe feared the loss of the company's small but profitable business in the state. But for the salesmen themselves the stakes were higher. Mississippi was a major front in an ongoing battle to change how textbooks were regulated across the United States. The new regulations might have had only a minor impact on the bottom lines of publishers, but it could have made the practice of sales agency—and Louis's and Nihart's jobs—obsolete.

Louis, Nihart, and two other agents had been sent by Howe to prevent Governor Theodore Bilbo from making good on a 1927 campaign promise to create a state printing plant to produce textbooks for the schools of Mississippi. If the plan succeeded, private textbook companies, like the ABC, would be shut out of the state's market.

Textbook reform was one of the major issues Theodore Bilbo rallied on in a populist campaign that had swept him back into the governor's office after losing reelection four years earlier. Textbook reform may seem incongruous, for Bilbo was not considered a great supporter of education. He was a polarizing, controversial figure in Mississippi politics throughout his life. An admitted member of the Ku Klux Klan who later in life became one of the staunchest Dixiecrats in the U.S. Senate, he saw his political career come to an abrupt end in 1947 when he was prevented from taking his Senate seat through the efforts of Ohio senator Robert A. Taft.[3] (In the late 1920s, prior to his election to public office, Taft had served as a Cincinnati-based member of the ABC's board of directors.) This is a political résumé that seems at odds with an image of Bilbo the education reformer.

In fact, Bilbo was not seen as much of a reformer despite his campaign against the textbook trust. He was heavily criticized just a few months after the 1927 election when he fired the chancellor of the state university and many members of the faculty, replacing them with political supporters.[4]

Nonetheless, his rhetoric and exhortation to the legislature "against the agents of the 'American Book Trust'" (as some rival publishers and school reformers called the ABC) had rung true with Mississippi voters. Bilbo quickly followed through on his campaign promises. Of a 109-page address to the legislature in early 1928, one-third was "devoted to the subject of a state printing plant for school text-books."[5]

The printing proposal, which passed the State Senate but failed in the House of Representatives, would have empowered the state to produce its own textbooks, which it would provide, at cost (including overhead), to its students. Louis and his colleagues' illicit lobbying and attempted bribery of Mississippi legislators in 1928 contributed to the defeat of the proposal. According to the Senate Committee that investigated them: "The American Book Company, having established an illegal lobby, beyond any question or doubt . . . is presumed under our law to have intended to use corrupt and fraudulent methods and practices in the carrying on of such lobby."[6]

Mississippi's printing proposal was one of many attempts by state governments across the United States to regulate the production and distribution of textbooks since the 1870s. Regulation proposals took on many forms, ranging from state-controlled printing to the negotiation of statewide prices for a period of several years at a time. Some states prescribed uniform adoptions, guaranteeing a market to publishers in exchange for lower prices and preventing local school boards from falling under the sway of textbook industry lobbying.

The period from the end of the U.S. Civil War to the 1920s was one of great changes in the way in which textbooks were produced, marketed, consumed, and regulated. Some of these changes can be traced to developments in pedagogy, but for the most part they came about because of new practices in the textbook industry itself and in American debates over business regulation. To understand the changing nature of textbook consumption, one must look at the companies that produced the books, the states and schools that regulated them, and the teachers and students who used them.

Those who are familiar with controversies over textbooks and their adoptions in recent years may be surprised by the fact that very little of the early history of state regulation of educational materials was concerned with matters of content. While some members of the textbook-producing community—mostly authors and editors—and many textbook consumers—teachers and students—focused on the textbook as a medium for conveying subject matter, state and local regulators most often focused on the textbook as a physical object that had to be produced, sold, distributed, and maintained.

So did the salesmen who were the producers' most direct contact with users. Sales agents often emphasized matters of price, availability, and durability. Most of the early efforts at regulation had to do with these qualities. Often, questions of pedagogical quality or content were only a secondary consideration.

One of the most important developments in textbook regulation was a shift away from local control to state-level adoption throughout much of the American South and West. This began even before the 1890 incorporation of the ABC and continued through the 1920s. This shift in who controlled adoptions ought to have provoked a response from publishers, but there was little change to the textbooks themselves. Statewide regulation came about as an effort first to control the behavior of textbook salesmen and then to curb the fears of price-fixing and monopoly that were associated with publishers' trade organizations in the late nineteenth century. In rare instances, regulation did affect the content of textbooks, but most of the publishers were slow to adapt their books' contents to political changes.

In 1928, the two primary reasons behind state regulation were still present in Mississippi: the high price of textbooks—driven by a "book trust" that kept prices inflated by monopoly or limited competition—and frustration with the practices of textbook salesmen, who were seen as insidious in their pursuit of profits and whose tactics seemed inherently corrupt. In the late nineteenth century, competition among individual salesmen working for rival companies had become so fierce that neither school boards nor publishers' boards of directors controlled the escalation of their numbers or tactics. This led to the creation of trade organizations within the schoolbook publishing industry. These efforts at self-regulation did little to stem the textbook salesman's growing notoriety or the industry's complicity in high prices.

The price of textbooks was an important political issue because in most places books were purchased directly by students (or more likely their parents). As school attendance became compulsory, this became an obligatory expense, which incited political opposition to mandatory public schooling. Individual schools and local districts tried all sorts of strategies to reduce the cost. In some areas, mostly northern urban school districts, the "free textbook" movement had already succeeded. This meant that a school board would purchase books directly and loan them free to students for the year, the cost of the textbooks being included in school funds that also paid teacher salaries and building maintenance. Free textbook schemes ensured greater book reuse, but they also shifted education expenses to school taxes.

Even before supporters of public education advocated free textbooks, they had already been pursuing methods of lowering the cost of books.[7]

By soliciting bids for a statewide adoption, states hoped to lower prices. That is, these states hoped to guarantee publishers a certain amount in sales, by limiting the ability of schools to choose alternative books, in exchange for lower overall prices. In addition to the lower listed prices for books, a fixed length of an adoption allowed for the resale or reuse of books over a multiyear contract.

Publishers fought against this effort, even going to court to protect their sales. Preventing book reuse was a large part of this. In 1901, the ABC sued a company that was taking damaged copies of its books, rebinding them, and selling them for less than the retail price. The company was subsequently required to label its rebound books but was permitted to continue selling them.[8] This was not the complete victory the ABC had hoped for, but its sales agents argued that rebound books were of inferior quality, and the labeling requirement made the rebound books clearly identifiable. In 1914, Ginn and Company (the second largest textbook publisher after the ABC) won a lawsuit that prevented rebinding companies from printing replacements of lost or torn pages so that they could sell more repaired used books.[9] By keeping used books out of circulation, textbook publishers could sell more new books at full prices.

Publishers also fought one another ruthlessly in their battle for sales, slandering their rivals, and instigating libel lawsuits to discourage criticism. In addition to legal tactics, individual sales agents (often with their company's central offices turning a blind eye) sometimes crossed the line into unlawful practices. They insinuated themselves onto local school boards, bribed elected officials, and arranged kickbacks with superintendents, teachers, and board members. Often, they would try to get school boards to change required books every year (or even more rapidly), limiting the amount of reuse, and compelling much more frequent purchases. In many parts of the country, sales agents constituted the only direct contact between consumers and the source of their textbooks, and the reputation their tactics earned came to stand for the companies they represented. Newspapers of the nineteenth century carried many stories of corrupt textbook salesmen. These antics became ready fodder for the advocates of centralized textbook control.

Supporters of state-level regulation hoped that reducing the frequency of adoptions, mandating fixed-term contracts, and placing adoptions under centralized control would prevent the worst of these practices from continuing. State control would guarantee longer periods when book reuse could occur, reduce the number of salesmen, and increase the credibility of the people responsible for choosing textbooks. By consolidating a state's adoptions into a single event every five or so years, it was even thought that state regulation might decrease the number of sales agents that companies employed (a circumstance that might have actually pleased publishers' boards of directors, though not the sales agents themselves).

Politicians invoked either the price of books or corruption in adoptions when advocating textbook regulation. An Illinois House subcommittee investigating textbooks in 1891 put forward a resolution declaring both that "it is the most universal belief of the people of this State that the prices of school books are exorbitant and that such prices are unjust" and that "such unjust and exorbitant prices so paid for school books are the result very largely of the combined and concerted action of certain persons . . . known as the 'book trust' or 'combine.'"[10] High prices and corrupt practices were interrelated. Corrupt relationships between book companies and local officials limited competition and allowed publishers to maintain higher prices; occasionally, publishers would use the promise of lower prices to sway elections. Some would offer to buy up used copies of competitors' books or take them as "even exchange" for their own. This practice enabled sales agents to use loss-leading textbooks to establish local monopolies, forcing some smaller publishers out of districts altogether or out of business entirely. Once an adoption was secure or rivals had conceded the region, the publishers could then set prices higher and recoup the expense of initially undercutting competition. The ABC's use of this tactic, charged rival publisher Edwin Ginn, had driven many smaller firms out of business altogether.[11]

These anticompetitive practices were the main reason that textbook publishers constituted one of the first major American industries against which accusations of trust were leveled. Yet in the history of American economics, schoolbooks are not often considered alongside oil, railroads, coal, and other major monopolies of the late nineteenth century (a perception perhaps reinforced, ironically, by the American history textbooks of today).

But was there really a book trust? When the state of Tennessee solicited bids for its 1925 textbook adoption, Commissioner of Education Harned identified fifty-five publishers, and more companies wrote him to request inclusion in the adoption.[12] There were probably more textbook publishers then than there are now in the United States as consolidation has led to only a few very large publishing houses serving the American textbook market. No publisher today, however, holds anything like the market share the ABC held at its peak in the 1920s. Most others were small publishers, focused on just a few subjects, such as spelling or drawing. Only a few large publishers had catalogs that covered most subjects. These larger companies fueled the competition for sales. Although it frequently denied that it constituted a book trust, the ABC sold about 80 percent of the textbooks in the United States. Many of the fiercest battles between it and Ginn escalated the prominence of textbook salesmen as a dominant force in the industry.

THE ASCENDANCY OF TEXTBOOK SALESMEN

In the antebellum United States, many textbooks were published by firms that also published books for general use. It was in this era that the idea of the textbook specifically for school use became more common (as schools themselves became more common in the United States). In some subjects, such as the sciences and history, books that were not originally intended for schools were adapted to that purpose. It was in this period that the notion of textbooks took hold. One of the first subjects for which books were written specifically for education was primary reading, which blossomed in the first half of the nineteenth century.[13] The marketing of books for school use caused concerns almost immediately. Massachusetts Board of Education superintendent Horace Mann wrote in 1837:

> Publishers often employ agents to hawk their books about the country; and I have known several instances where such a peddler,—or picaroon;—has taken all the old books of a whole class in school, in exchange for his new ones, book for book,—looking of course, to his chance of making sales after the book has been established in the school, for reimbursement and profits; so that at last, the children have to pay for what they supposed was given them. On this subject, too, cannot the mature views of competent and disinterested men, residing, respectively in all parts of the state, be the means of effecting a much needed reform?[14]

Mann offered one of the earliest expressions of concern over the use and marketing of textbooks; such concerns would only increase after the Civil War.

As the number of students multiplied and the amount spent on textbooks grew exponentially, publishers began to employ strategies to assure themselves sales in an increasingly competitive environment. As John Tebbel's history of American publishing relates: "By this time, specific texts were likely to have lodged themselves solidly in a school. Citizens who lived in the town and served on the school boards had been brought up on them, and were not inclined to adopt others, although they felt increasing pressure to do so as the number of textbook publishers increased and competition intensified."[15] By the 1850s, textbook publishers employed traveling agents to lobby school boards and to advertise their books.

This sales model was also necessary because many books being sold for schools in the early period of this history were anthologies composed of classic works of literature, not works written exclusively for the publisher. In the era before international copyright protection existed in the United States, many of these classics (and even contemporary works from abroad) were available from multiple publishers. There was little reason to emphasize a book's *content* as an exclusive selling point. The cost of the books and their physical qualities—such as the clarity of their printing and strength of their binding—coupled with the personal charisma and salesmanship of a company's representatives were the primary bases for choosing one schoolbook over another. The lessened emphasis on textbook content was exacerbated by the organization of sales agency by geographic region, not by school discipline. A salesman who had a relationship with a superintendent could as easily sell a math book as a history book. Rarely would a salesman's knowledge of the inside of a book matter.

After the Civil War, textbook control became a much larger issue. In addition to Mann's concerns over sales practices, there had also been some controversy about the content of textbooks, most notably U.S. history books' treatment of slavery.[16] Additional concerns emerged along with newly developed structures for printing, distributing, and selling books. Improved printing and transportation technologies (such as linotype printing and more efficient railroad coverage) made it possible for publishers to produce more copies and distribute then more widely. Publishers began to expand to larger territories and hire more sales agents to canvass new areas. Publishers from different cities were more likely to find themselves in direct competition.

A new culture of professional salesmanship emerged throughout American industry. This new commercial environment placed greater emphasis on a salesman's personality and his relationships with individual clients than on the qualities of the particular items being sold.

As textbook companies competed in this industrialized and professionally transformed environment, Horace Mann's early call for state-level reform gained new resonance; it seemed unlikely that local regulation could contain the expanding textbook industry. New companies formed as partners or employees of the older publishing houses set off on their own.

The example of Ginn and Company, which began in this way, is illustrative. A history of the company written by one of its employees in 1938 compares its founder, Edwin Ginn (1838–1914), to Horace's description of early ocean explorers, who "must have had a heart of brass and a nerve of steel": "We might well think Edwin Ginn must have been similarly endowed." This heroic account depicted a man who took a great risk in going into business for himself in educational publishing in the aftermath of the Civil War: "Besides the usual five senses given to man, he had two more—common sense and vision. He saw visions, and he dreamed dreams, and he wished to have a business of his own."[17] Ginn is also cast as a public servant, and the job of textbook publishing is described as a noble calling.

Edwin Ginn entered the textbook industry in 1858, working as a salesman with a Boston-based publisher. He succeeded in winning the adoption of his employer's textbooks in that city and other nearby districts. No mention is given of what sort of sales tactics he employed in those days, but he undoubtedly benefited from the large student population in the area. Owing in part to Horace Mann's efforts, Massachusetts had been one of the first states to enact compulsory public education laws. In 1867, Ginn went into business for himself by buying the plates of one of the texts he had previously sold: George L. Craik's *The English of Shakespeare*. This work, which interpreted passages of Shakespearean verse with explanations of the text and vocabulary, had first been published in England, and its copyrights were not protected in the United States.[18] When Harvard University adopted the book in 1869, Ginn's immediate financial success was assured.[19]

Ginn rapidly expanded his business, publishing a variety of classic works of literature, and expanding his company's reach by hiring more sales agents. This expansion was facilitated by the growth of compulsory schooling in other states and eventually brought the company into direct competition with other major publishers outside New England.

FROM TRADE ASSOCIATIONS TO TEXTBOOK MONOPOLY

In the late nineteenth century, competition between publishers—and the cost of that competition, reflected in the increased expense of sales agency—was reaching unsustainable levels. Publishers became more dependent on their salesmen as they expanded. Rarely did editors or executives travel to local markets. Reports of what was wanted by schools and needed to secure adoptions came from sales agents. Often, these reports described the great number of rival agents against whom salesmen had to compete. These men had jobs tied to their performance. Arguing that their own numbers were unnecessary or that any failures were due to their own faulty efforts rather than a lack of support from the company would not have been in their own interest. As companies sent more agents to a region, others responded in kind, leading to an arms race of salesmen that became inherent in the industrial organization of textbook companies. Despite nearly fifty years' worth of attempts by publishers to address this condition, the outsized influence of textbook salesmen on their companies persisted into the twentieth century. As late as 1926, Ginn and Company president George A. Plimpton noted that sales agency expenses continued to increase faster than editorial expenses, even at a time when the company's overall sales were falling.[20] In effect, even though they were rivals, salesmen for different companies worked together to keep their respective bosses devoted to the importance of the agency strategy.

More expensive sales agency began cutting into publishers' profits in the late nineteenth century. Companies became hostages to their own competition. A publisher that tried to unilaterally disarm by reducing its sales force would be wiped out, but to continue to escalate sales efforts threatened to make the publishing enterprise unprofitable (or else make textbooks even more expensive, which further inflamed relationships with consumers). It became clear to many publishers that, if they did not reform themselves, states would do it for them. This led to the creation of trade associations.

In 1870, Ginn joined in the formation of the Publishers Board of Trade, initiated by John C. Barnes, of A. S. Barnes and Company.[21] The Publishers Board of Trade established bylaws meant to reduce the expense of competition by limiting the amount of discounts permitted to school boards and reducing the number of salesmen. So little was the trust between competitors that it even legislated public disclosure of the identity of members' salesmen so that no company could hire secret agents. This would also prevent salesmen from running for a school board without acknowledging their affiliation. However, the failure to get many publishers to join and the impossibility of enforcing obedience among those that did quickly led to the undermining of the organization. An unpublished history of the ABC by one of its employees describes this: "These hold outs, the continued watering down of the By-laws, and the realization that many members were professing reform and practicing the same old abuses, had its effect on John C. Barnes."[22] Barnes retired in 1873. Shortly afterward, some of the largest publishers withdrew, and the Publishers Board of Trade collapsed in 1877.

The ABC history continues: "Out of this attempt, however, came a new concept, the textbook publisher's *syndicate*, a term which later became one of approbation and was promptly changed to *alliance*." In 1880, the three largest educational publishers entered into the Publishers' Alliance: Van Antwerp, Bragg and Company of Cincinnati, publishers of the *McGuffey Readers*; Ivision, Blakeman and Company of New York, whose catalog included Asa Gray's botanical writings; and A. S. Barnes and Company, also of New York City. Though the three companies continued to function separately, they agreed not to compete against one another: "None of the three houses would 'seek to displace the books of the others, directly or through agents.'" These three were joined in 1881 by the fourth-largest schoolbook publisher, D. Appleton and Company (which was also the American publisher of Darwin's *Origin of Species*). In 1885, this alliance grew into the School Book Publishers' Association, after an agreement was signed by sixteen publishers, including the "Big Four" and Ginn.[23]

As the association encompassed more companies, concerns about monopoly grew. An 1889 article in *Publishers' Weekly* addressed this:

> Some time ago some of the leading school-book publishers formed themselves into an association with the purpose of working in harmony rather than in opposition, that is to say, of restraining their rivalry within specific limits. . . . It is not, in any sense a Trust, nor will it destroy competition of the right sort. Any business arrangement which undertakes to diminish competition in quality or price is against the public interest and ought not to succeed. This combination, as we understand it, has the contrary purpose of serving the public by saving labor and cost. If it should go farther than this, and bring the trade into the stagnation which comes from the destruction of competition, the results would not be good, and the arrangement, we believe, would not last. There is not the slightest thought, however, of any such scheme.[24]

Despite the reassurances of *Publishers' Weekly*, in 1890 the School Book Publishers' Association was dissolved, and the four largest publishers agreed to consolidate and incorporate, forming a new entity, the ABC. According to the ABC's internal history: "There is some evidence that an attempt was made to draw other houses into the group, in particular Harper & Brothers, Ginn & Co., and D. C. Heath & Co."[25] All those other houses refused.

Though some considered the new corporation a trust, others were less convinced. *Publishers' Weekly* continued to defend the new ABC, observing: "The combination includes the four leading firms of the school book trade, doing probably no more than half the business; there are between one and two hundred other concerns in the trade, some of them of large individual importance."[26] ABC officials also brushed off the allegations. "As business men, fully aware of the odium which attaches to the name of 'trust' or 'monopoly' in the public estimation," one ABC executive stated, "we could not afford to do anything that would justify the application of the term to our organization. To do so would be suicidal and, of course, we are not proposing to do anything suicidal."[27]

But just a few weeks later the ABC stunned the industry by purchasing the textbooks of the fifth (or, after the consolidation, second) largest educational publisher, Harper and Brothers, which had just refused to join the companies forming the ABC. This was not the only surprise. "The American Book Company, by some magical means, has succeeded in accomplishing what few would have believed possible three weeks ago—the control of the public school books of Harper & Brothers, and of the plant of The Standard Publishing Co. and D. D. Merrill & Co.," *Publishers' Weekly* reported. "Their surrender and the purchase of the Harper books now practically puts all the public school business in the hands of the American Book Company, and places it in a position of grave responsibility."[28]

Though *Publishers' Weekly* made no mention of the connection, the "magical means" by which Harper's textbooks came to be sold was the unexpected death, on May 22, 1890, of Fletcher Harper Jr. He was the largest stakeholder in the company and had designated no heir. Faced with amortizing his equity, the day after his death the Harper board held a meeting at which fears over the company's available capital were revealed. Accepting the ABC's offer had suddenly become both reasonable and necessary.[29] The company received "$400,000 for the rights and an additional $150,000 to cover plates, sheets, and bound stock."[30] If not for Fletcher Harper's unexpected death and the company's sudden need to raise over half a million dollars cash, the ABC might have had a second strong rival along with Ginn.

The need for publishers to consolidate and to rethink some of their practices in this period may also have been motivated by the growing sentiment in the United States in favor of the recognition of international copyrights. Debates over the extension of copyright protections to non-American authors had built to a crescendo, culminating in legislation in 1891.[31] This had a unique importance for textbook publishers, who incorporated part or all of foreign (mostly British) works into readers and other textbooks. New copyright laws meant that they would have to develop different relationships with their authors.

The consolidation of many of the leading schoolbook publishers into a single corporation did not bring an end to corrupt practices in the industry. The first test of the ABC's new power was the textbook adoption in the state of Washington in June 1890. When the ABC lost the initial vote on math books to Ginn, its sales agents tried some old-fashioned techniques to promote the new corporation, techniques that unexpectedly backfired: "According to the prosecuting attorney, the American Book Company agents, S. W. Womack and R. L. Edwards (already famous for Montana scandals), 'did then and there, unlawfully, wickedly, and corruptly' pay one of the adopting board, L. H. Leach, $5000 *by check* to reopen the adoption. An indictment was handed down."[32] In the course of the state's investigation, it came to light that the agents' motivation for bribing the secretary of the state board of education was the suspicion "that Leach had received $2,500 from other concerns to vote for the adoption of their books."[33] According to the *Chicago Tribune*, the ABC agents hired a private detective from Portland to "test" Leach by giving him the check in exchange for reconsideration of the adoption. Criminal charges were brought against the two ABC agents, but they fled to British Columbia before they could be arrested.[34]

Later that year in Mississippi, "certain persons, acting as agents of the American Book Company, were going from school to school in counties where the books of the counties had not been adopted, and were inciting the people to refuse to supply their children with the adopted books."[35] Though the incorporation of the ABC changed many aspects of textbook marketing, some old tricks remained the same.

The formation of the ABC and its purchase of Harper's schoolbooks left Ginn as the next largest textbook publisher, and its founder set out to portray himself as the champion of the schools against the corrupt monopoly. In 1891, Edwin Ginn wrote an essay published in the *Saturday Evening Post* addressing the behavior of the ABC. He republished this in 1895 as a pamphlet entitled *Are Our Schools in Danger? The Great School-Book Combination* for his sales force to distribute. In this essay, Ginn argued that the ABC's very existence was detrimental to schools, to students, and to school boards. He particularly decried even exchange. Smaller houses could not afford to emulate this, and the practice helped the ABC expand its domination of the field. He warned that districts opting for even exchange might suffer in the long term, when the ABC was the only company left:

> We have given the best years of our life to trying to produce good books and place them on the market, by honorable methods, at a fair price. We must leave the results of our efforts to the judgment of others. . . .

> Should the American Book Company by even exchanging the books of their competitors in the schools narrow down the holdings of other houses so that it will be impossible for them to continue business at a profit, it is easy to see that the day when these houses will have to relinquish the field will not be far distant. When the American Book Company have the entire business in their own hands, it may be natural for them to sit down and calculate how much it has cost to get rid of their competitors, and say to themselves, "Now it is only fair that the public, who have had the benefit of these free books so long, should pay for them."[36]

The pamphlet brought greater attention to the Ginn-ABC rivalry and refocused interest on the question of whether the ABC was a monopoly. The ABC responded with its own eight-page circular dated June 10, 1895, entitled *Who Is Edwin Ginn?* In it the ABC claimed that Ginn really was a pretend friend to the education establishment and that, like any businessman, he was concerned only with his own profits: "In substance this Ginn pamphlet says: I, Edwin Ginn, know all that is knowable about the schoolbook business, not only about my own business, but about everybody else's business, and have appointed myself the supreme judge of publishers and the guardian of teachers and school officers." The ABC also decried the accusation that it constituted a monopoly. Observing that Ginn and Company had distributed its pamphlet throughout Vermont prior to the 1895 adoption, the ABC rebutted: "Nobody knows this to be false better than Ginn & Company do. Nothing proves it to be false more completely than the Vermont canvass itself. . . . The American Book Company had 15 agents only employed in the Vermont canvass. It is reported that Ginn & Company had there 25 agents. Other houses must have had the balance of the 175. Competition and not monopoly would seem to be firmly established in the schoolbook trade, notwithstanding Mr. Ginn's great and continuous wail to the contrary."[37]

The Ginn-ABC textbook battles continued throughout the 1890s. Other publishers were drawn into the conflict. In 1898, the ABC sued the Kingdom Publishing Company of Minneapolis for libel. It had published a pamphlet entitled *A Foe to American Schools*, written by George A. Gates, the president of Grinnell College.[38] The ABC alleged that Gates was acting on behalf of Ginn in falsely accusing it "of being a trust and monopoly, of being illegal in its organization, of being unscrupulous in its business methods, of securing the elections of trustees favorable to it by intrigue, bribery, threats and promises, of suborning newspapers, and dishonestly and unlawfully silencing opposition."[39] It boasted of its "vindication" in a pamphlet of its own.[40] This was sent out to state and local education officials accompanied by a cover letter stating: "1000 copies, comprising one-half of the first edition of this pamphlet, were furnished by the Kingdom Company to Ginn Company, Chicago, by order of President Gates." The letter closed with the pronouncement: "We shall hold to legal accountability all parties circulating this pamphlet or any similar libel against us."[41] The Kingdom Publishing Company was put out of business by the lawsuit, a chilling warning to publishers considering challenging the ABC.

The perception that the ABC was a monopoly—encouraged by Ginn and perhaps borne out by its market dominance—was the basis for authorities in Texas to exclude it from state adoptions in 1897. This exclusion was not reversed until 1924, as the ABC's history relates: "Somewhere in our files there is the photograph of a freight train bearing a long streamer proclaiming that the train was entirely filled with geographies on the way to Texas. We had at last written an epitaph to the ABC-Ginn war."[42]

The end to the ABC-Ginn war was most certainly helped along by the death, in 1914, of Edwin Ginn. The mantle of leadership of his company passed to George A. Plimpton, who shared many of Ginn's political and philanthropic concerns, but who had experienced less of the bitter acrimony dating back to the nineteenth century. Plimpton had joined Ginn in 1881; his background was editorial rather than sales. This may explain why he was less emotionally invested in the Ginn-ABC rivalry and why he was able to be critical of his own sales force in the 1920s.

The ABC was accused of a list of crimes that may not have been complete or accurate, although the evidence of several cases of illegality was well documented. The 1928 Mississippi case shows the ABC engaged in corruption, attempted bribery, and misrepresentation. But it is unlikely that it was unique in this behavior. Was the ABC a trust? Whether or not it was ever legally a trust, as it was judged to be in Texas, it was the frequent target of those who saw monopoly and corruption in the textbook industry.

Throughout the 1890s, Ginn accused the ABC of anticompetitive practices, and the ABC accused Ginn of hypercompetitive practices. Both accusations played into the debate over state-level regulation of the textbook industry. Both Ginn and the ABC agreed that more regulation was undesirable. In *Are Our Schools in Danger?* Edwin Ginn claimed: "I tried very hard to persuade [the companies consolidating] not to form this company, for I feared that it would tend to stir up fresh legislation in regard to the book business, which has generally been an injury to both publishers and the public."[43]

Attempts to regulate the textbook industry gained momentum shortly after the consolidation of the ABC in 1890. Local governments were ill equipped to control large corporations, and the federal government never addressed the textbook industry among the other great trusts of the period. It is likely that textbook regulation was regarded as an education issue, for states to govern, rather than an interstate commerce issue, under the jurisdiction of the federal government. Despite this, the rhetoric that grounded textbook regulation was less about textbook content and more about sales practices. Nowhere, perhaps, was this concern about sales practices more apparent than in California, where textbook reform was seen as crucial to solving the bachelor problem.

TEXTBOOKS AND THE SINGLE MAN

When California considered a new state constitution in 1879, textbook publishing took up a significant amount of the debate. State support of education was seen as vital to California's growth and the diversification of its economy. The gold rush had attracted to the state mostly unmarried men, and a greater emphasis on education would, it was hoped, lead to a more balanced population.[44] "We can create no fund too large for the purpose of education," argued John McDougal, a delegate to the constitutional convention from Sacramento. "I call upon my old bachelor friends to support this if they want wives, for it will introduce families into this country."[45] Of the more than sixty-five thousand people aged sixteen to fifty-nine living in San Francisco in 1870, men outnumbered women by more than six to one.[46]

The manner in which textbooks were regulated had little to do with the surplus of bachelors in California, but it had everything to do with the twinned concerns of monopoly and corrupt textbook salesmen. The convention delegate James O'Sullivan proposed imposing the four-year minimum on adoptions to get "rid of the infamous lobby around the legislature." Though he lost in his effort to have the state control adoptions, the local boards were required to have four-year adoptions.[47]

O'Sullivan also proposed an additional section to the constitutional article on education: "The Legislature, at the first session after the adoption of this Constitution, shall provide for the appointment of a Commission to compile a series of school text-books, which, after preparation, shall be printed in the State Printing Office. The text-books thus compiled and printed shall constitute a uniform series of text-books, to be used in the public schools of this State on and after the first day of January, eighteen hundred and eighty-two, and shall be furnished to pupils at cost price." In support of this, O'Sullivan offered two reasons: "First, the adoption of the system which I propose will completely destroy all motive or inducement to a lobby on the school book question. Once the State prepares and publishes its own school books, the corrupting influence attached to the private competition and speculation of the present system will be effectually prevented in the future." He continued on the effects of the textbook lobby: "These lobbies were here in force about the last Legislature, and if it were not for their influence, it is not improbable that this school book question would have been settled at that session." When the proposed section was defeated, O'Sullivan castigated his fellow delegates: "At the last session of the Legislature, bills were introduced . . . proposing to take this vexed question away from the Legislature, and away forever from the corrupt lobby influence. But Bancroft had his agents, attorneys and lobbyists upon the floors of your halls, and the bills never became laws. McGuffey and Bancroft are powerful in this building. I understand that both firms have their legions hovering around this Convention. Some of our teachers are simply agents and lobbyists for Bancroft & Co."[48]

State regulation was defeated in California in 1879, but the debate itself shows quite clearly what its supporters thought mattered. Textbook content was *never* an issue in this discussion. The cost of books and the effect of corrupt textbook lobbies were. In 1879, these concerns were not able to overcome skepticism of centralized control. However, problems with textbook publishers had grown great enough just five years later that state printing was established by an 1884 amendment to the constitution.[49] The 1884 amendment had the support of all major political parties and had been endorsed in each of their state platforms.[50]

From California, the issue of textbook reform spread to states with more balanced gender ratios. In Indiana, frustration with the corrupt actions of sales agents culminated in a populist decrial of monopoly in 1889. As one contemporary account describes, many local newspapers added their voices to the cause of reform: "Patriotism and revenge worked side by side in raising in the ready press the war-cries: 'Smash the book trust!'; 'Cheap books for the children!'"[51] This led to the passage of a uniform textbook law. It was the start of a major trend—a *Publishers' Weekly* article noted that "an unprecedented number of bills" regarding the provision and adoption of textbooks had been introduced in state legislatures "within the last three years," that is, since 1892. By that year, thirteen states besides California had their textbooks selected at the state level.[52] In 1897, the Kansas legislature passed legislation creating a textbook commission and requiring statewide uniform adoption, and, in 1913, Kansas became the only state besides California to require the state printing of textbooks.[53] That same year, Georgia, which already adopted textbooks at the state level, appointed a commission to investigate the possibility of printing textbooks.[54]

In 1899, Tennessee passed its first uniform textbook law, whereby a state textbook commission solicited sealed bids from and entered into five-year contracts with publishers. Textbooks were then distributed to authorized regional depositories, which in turn were expected to provide an adequate supply of books to local retailers.[55] Tennessee was one of the first southern states to adopt textbooks statewide, but eventually state-level control became the law across the South. Several southern states (e.g., Mississippi in 1928) also looked into the feasibility of state publication, though none went through with it.[56]

The ABC had had no business in Texas for years; reflection on its return to the state showed the company's ambiguous reaction to statewide adoption: "This success, however, proved that the single-basal state adoption was a decidedly mixed blessing. More than anything else it turned the thoughts of a number of our key executives backward nostalgically to our great days and prevented them from planning adequately for the new era rapidly developing."[57] While such a system could potentially benefit publishers by guaranteeing longer adoption periods and opening entire states to sales all at once, publishers lobbied against state-level adoption. In some cases, the practice may have been seen as leading to state publication strategies like California's, but the main reason for their objection to it may have had more to do with the role of sales agents in the industry's organization and less to do with any actual risk of lost textbook sales. State adoption could succeed where trade associations had failed, reducing the number of salesmen by making the agency strategy obsolete. With fewer (albeit bigger) adoptions, fewer agents would be needed. It was these agents, often men without families who spent many of their days on the road, who also reported back to the publishers' main offices on local conditions. They knew that state regulation would undermine their jobs and described the regulatory trend as a threat to their companies, not just to themselves. The industry's reaction to state adoption was unsurprising.

Executives and editors might have taken notice of statewide adoption more quickly had it occurred in New York, Illinois, Pennsylvania, Massachusetts, or Ohio, where nearly all of them were based. Another form of regulation consisted of a state board of education fixing maximum prices for books while leaving to local boards the option to choose among many books. Ohio enacted a plan like this by 1905, requiring local boards to adopt books for five years.[58]

Many of the states that did *not* have statewide adoption were also those with the largest numbers of students. In 1870, only four states had school enrollments over 500,000: New York, Pennsylvania, Ohio, and Illinois. By 1920, the overall number of students enrolled in schools was much higher, yet the largest school populations were still in New York, Pennsylvania, Illinois, and Ohio. These were the only states with more than one million students enrolled. Massachusetts was not far behind. Despite the spread of state regulation, about 68 percent of students in the United States in 1920 attended schools in states *without* statewide adoption.[59]

Some of the states that did not enact a system of statewide uniformity were those most greatly affected by the violence that occurred in the mid-nineteenth century over the use of the King James Bible in compulsory public schools. The so-called Bible Wars, which led to riots in Philadelphia and Cincinnati and unrest in many northeastern states, also contributed to the rise of separate Catholic school systems.[60] Many Bible War cities were also major publishing centers, and their influence might have had more to do with why the states in which they were located did not pass uniform textbook adoption laws. The legacy of religious and anti-immigration violence over using a book in schools may not have been a major reason that state-level control did not come to pass in the northeastern and midwestern United States, but the Bible Wars likely did influence how other states regarded the use of the Bible in schools. State constitutional conventions in Colorado in 1875–76 and in California in 1879 prevented the use of the Bible as a textbook, but debates in both these states framed the prohibition as a matter of religious freedom, not one of education control.

THE CALIFORNIA PLAN BACKFIRES

The California publishing plan seemed like the perfect solution for the rapidly growing state. Printing its own textbooks would save the state a tremendous expense. Meddlesome textbook agents would be gone. The state could even encourage home industry by hiring faculty at the new universities to write the state's textbooks. In practice, the plan quickly fell apart. In 1885, the legislature appropriated $150,000 for the expense of equipment and manufacturing and an additional $20,000 for the printing of textbooks. After the initial start-up, state printing was supposed to pay for itself, funded by the books being sold at cost. This proved to be unrealistic; between 1887 and 1911, over a half million dollars more had to be allocated to the printing office.[61]

When James O'Sullivan proposed his original textbook printing amendment in 1879, one of the arguments he advanced was that it would encourage local authorship. This was attempted with mixed success; the state board of education hired teachers and university faculty to write educational materials. But the effort to procure sufficient and qualified writing talent was flawed. The best California authors were offered greater royalties by commercial publishers. Books being written in California were of inferior quality. And, more importantly, schoolteachers and the general public *perceived* the books to be inferior to national textbooks.[62] In 1890, a resolution was adopted at the convention of California school superintendents stating: "Certain of the state text-books . . . have met with the approbation of the public school teachers of the state, we desire to record our severe criticism and disapproval of others of the state series."[63]

By 1903, the state abandoned local authorship. It authorized the state printer to lease existing plates from commercial publishers and use them to print books.[64] But, because this was allowed, much of the original purpose of the constitutional requirement was lost. The practice of seeking plates from commercial publishers reintroduced the potentially corrupting influence of the textbook industry. The state still printed and bound books, but it did not alter their content. Paying to lease the plates also negated much of the planned savings in cost. Predictably, many publishers, including the ABC, refused to lease plates at all, expecting that they could soon starve the state plant out of existence. Nearly half the textbooks in the United States were unavailable in California because their publishers, including most of the largest, refused to lease plates for statewide adoption.[65] The companies that did allow their plates to be leased frequently did so only for books that were already several years old and were no longer being produced. A primer originally published by Ginn in 1891 was printed in California in 1905.[66] The books were identical except for the cover and the title page, and the original author was no longer credited.

It was hard for California to obtain the rights to new materials. Once they were obtained, it frequently took over a year to actually distribute new books to the schools. Textbooks often stayed in circulation for a dozen years and in some cases as many as sixteen.[67] California might not get the plates to a book until it was already off the market in the rest of the United States.

Most textbook publishers sought to replace or update textbooks within ten years of their introduction. It was possible that a child in California might have been using a textbook that was first written a quarter century earlier. This was obviously not what James O'Sullivan and others who supported his proposal in 1879 had in mind.

California's state monopoly on textbook printing also demonstrated that private commercial publishers had no monopoly on corruption. In January 1913, the state printer was condemned for his handling of his duties. A Senate committee reported: "The manner in which the State Printing Office was being and had for years been conducted . . . was as deplorable as it was astounding, and . . . tolerated a system reeking with fraud and dishonesty."[68] Graft and corruption as well as honest inefficiency cost the state far more money than it likely could ever have saved by printing.

Beyond corruption and inefficiency, the California printing amendment was criticized because of its effect on textbook content. California printed obsolete books, and not necessarily the best of those, but only the best of those that it could lease. While no one accused the state of having an ideological agenda in its selection of content, publishing companies helped cultivate a perception (through pamphlets and advertisements) that it would never have as up-to-date textbooks or books that were as good as the rest of the country's. Even though some texts, like basal readers that included passages of classic literature, could not really be said to have obsolete content, publishers also emphasized improved illustrations and the use of the latest pedagogical theories to promote their new books.

The trend from 1879 to the 1920s toward state-level regulation of textbooks was not explicitly about controlling content; nonetheless, content was affected. Publishers who had been shut out of California claimed that the content of state-printed books was inferior to that of the books they had for sale. California was the first state to regulate textbooks at a state level; it was also the first to suffer from unintended consequences of that regulation.

TEXTBOOK CONTENT AND UNINTENDED CONSEQUENCES OF STATEWIDE REGULATION

Over the course of the half century between California's constitutional convention and the Mississippi state printing bribery scandal, there were two general trends in the regulation of educational texts. One was the replacement of local control with state-level regulation, through either state printing or uniform adoption. The other was the focus on controlling prices, book durability, and textbook industry corruption as the motivation for increased regulation. There was little to no attempt either to justify state-level regulation or to criticize it on the basis of its potential to control textbook content alone. Content was one factor considered in individual adoptions, but no one seemed to think that the role of content in the adoption process would change under different regulating bodies. If anything, anticorruption advocates preferred that textbook selections be based on quantifiable measures—such as price, paper thickness, and cover durability—instead of allowing the too-subjective invocation of content quality to mask illicit influences.

Textbook publishers also made few changes to their books as a result of state regulation. If anything, the reaction of most publishers was simply to oppose state regulation. The ABC's reaction to Bilbo's state printing proposal was an extreme example of this. But the company's agents in Mississippi were coordinated, not by its corporate offices in New York, but by W. T. H. Howe in the Cincinnati office, which handled the South—the region where most state-level regulation occurred. Other companies that were more centrally managed had even less of a response to the phenomenon.

State boards were thought to face greater public scrutiny and were expected to be a check against corruption. They also made greater use of expert opinions than did local boards. The use of experts was often meant to ensure transparency in adoptions. It also led to a greater focus on content in the textbook adoption process. This also resulted in the influence of education journals and of the testimonies of professors of education at places such as Teachers College in New York, the University of Chicago, the University of Iowa, and—especially in the South—Peabody College in Nashville. Unsurprisingly, professors of education subscribed to a belief that schooling ought to be widespread, and many of these experts were strongly influenced by the progressive understanding of education put forward by John Dewey. The support of these professionals, or their opposition, could strongly influence the outcome of an adoption, especially when one of their fellow faculty members was a textbook author. Publishers and sales agents sought endorsements as well as feedback from them.

Perhaps more than anyone else involved with textbook marketing, the ABC managing partner, W. T. H. Howe, recognized the importance of pedagogy experts in the statewide adoption process and urged the company to take advantage of it. In March 1924, he wrote to the New York offices: "The University of Iowa controlled the State adoption, practically as far as educational sentiment goes, in the last State adoption in Louisiana." He continued: "One of the authors of Macmillan's geography is in Peabody. That is what I believe what we might call the Columbia of the South and they are going to make things very difficult for us."[69] The response, from the assistant editor in chief (the editor being on leave owing to an injury), suggested the company's strategy: "Of course, I realized that Peabody probably had a great sphere of influence throughout the South. I think we ought to get hold of some of those progressive fellows in that institution and tie them up to use one way or another because I imagine that Peabody is slated for a long and useful career in the world of pedagogics and especially in the South."[70]

THE ABC AND THE SOUTH: TENNESSEE AND THE SCOPES TRIAL

This ABC strategy of "t[ying] up" experts in order to use their influence is one of several historical connections between the ABC and the 1925 Scopes trial. After the ABC-published *Civic Biology* (which had been used by Scopes) had been tarnished by its association with the trial, Howe solicited feedback on the revised manuscript from the Peabody College biology professor Jesse M. Shaver. Shaver later wrote a positive review of the resulting book, the 1926 *New Civic Biology*, in the *Peabody Journal of Education*.[71]

In part, the Tennessee antievolution law was a result of the state's laws regarding statewide textbook adoption (passed partly in response to the formation of the ABC). The state's textbook commission was compelled to choose between biology textbooks that were pitched toward urban school and those for rural schools. Choosing the *Civic Biology* exacerbated that division within the state.

But it seems that the connections between the ABC and Tennessee's antievolution law may run even deeper. Evidence of this does not come from Tennessee in 1925 but from the Edwards Hotel in 1928. W. F. Nihart had reassured Howe that he and his fellow agents were "doing all that is humanly possible." But, back in Cincinnati, Howe was not convinced. On March 25, ten days after Howe had received Nihart's reassurances, the agents received a letter containing instructions, including advice to hire more help. Howe's impatience with the agents' progress in Mississippi was matched by their frustration with his management. These sales agents were used to having a great deal of autonomy and, as long as they got results, not having their methods questioned. Yet they feared that perhaps Howe himself had hired the detectives they were beginning to suspect were following them.

Lying on the floor of the adjacent room, the Pinkerton detective Joseph Littlejohn, who had been hired by Governor Bilbo, listened to these speculations. Through the wall, he heard Henry Louis, the lead ABC agent in Jackson, explain: "Howe doesn't know the situation here as I do, for I am on the job constantly. If he will just tell me to go down and handle it to suit myself, as I did in Tennessee, I could work better. You know I defeated the bill in Tennessee in the House almost unanimously."[72]

Louis as much as admitted that he had already done in Tennessee what he was trying to accomplish in Mississippi. That ABC agents made a habit of fighting state printing proposals is not surprising. The previous day, Louis had been overheard discussing his effectiveness in another legislative battle: "I can write the bill. I went to California and wrote the bill myself."[73] What is striking—and raises serious questions—is the fact that the Tennessee legislature considered state printing of textbooks in March 1925 during the same weeks that it considered and passed its antievolution bill.

A Tennessee House subcommittee investigating the question of printing textbooks reported that it had held hearings and invited "publishers of textbooks, editors, printers, and other persons familiar with the conditions relative to the prices and character of textbooks used in the public schools of Tennessee." As a result of these hearings, the subcommittee concluded: "The cost of plants and the small volume of work would result in a heavy deficit each year if the State should undertake the publication of its own books. . . . [W]e would recommend that the Governor and the Textbook Commission yet to be appointed use every means at their disposal and all their energies to procure for the children of Tennessee cheaper schoolbooks but not inferior books."[74] Tennessee decided against a state printing plant. In the overheard conversation in Mississippi, Louis admitted that the textbook legislation that emerged, which rejected state printing and also made distribution of textbooks easier for out-of-state publishers,[75] was largely the work of his lobbying. But this was the same week that the legislature had sent the antievolution bill to Governor Peay for his signature. The biology textbook in use throughout the state's high schools was the ABC-published *Civic Biology*. Henry Louis, the Nashville-based agent who was taking personal credit for defeating state printing in Tennessee, was also the ABC agent who had presented the *Civic Biology* to the state textbook commission back in 1919.[76] It was his book, his adoption, that was most directly affected by the antievolution law.

The antievolution bill was neither defeated nor vetoed. From this, it might appear that, even if Louis had tried to influence the bill, he was unsuccessful. But there are other ways to measure success, and the Tennessee antievolution law was significantly different from laws that had been passed or considered in other states. Oklahoma's short-lived antievolution law had outlawed the use of *textbooks* that included evolution.[77] Governor Peay even took the unusual step of sending a letter to the legislature (accompanying the signed bill) in which he stated, in part: "After a careful examination, I can find nothing of consequence in the books now being taught in our schools with which this bill will interfere in the slightest manner."[78] Instead of merely signing the bill, Peay went out of his way to implicitly exonerate the ABC's textbook. This statement from the governor could even be used by ABC salesmen to convince wary antievolutionists in Tennessee and elsewhere that Hunter's *Civic Biology* was unobjectionable. Given the possible forms an antievolution law could have taken, Tennessee's law was the best possible outcome for the vendors of the *Civic Biology*. But this phrasing also made it possible to try individual teachers like John Scopes for violating the antievolution law.

So did the ABC's sales agents have a role in the creation of the antievolution law? Perhaps they unsuccessfully worked to defeat the bill, or perhaps they were successful in having the bill regulate teachers instead of explicitly banning their textbook. Governor Peay's statement was effectively an endorsement of Hunter's book. Peay and Education Commissioner Perry L. Harned were both wary of textbook publishers holding political influence, so it is unlikely that they were directly lobbied by Louis. Nonetheless, the way in which the antievolution law was written, passed, and explained was a kind of success for those who stood to profit from the continued use of Hunter's *Civic Biology*.

If this kind of evidence were presented in the kind of hard-boiled detective story that the book agents' behavior fits so well, the formula of motive, method, and opportunity would be all one would need to indict the ABC. History, however, demands a greater burden of proof. Unlike detective fiction, it does not require any one party to be fully culpable in order to be held responsible for past events. Even if the ABC, Howe, Louis, and others were in no *direct* way responsible for the Scopes trial, there is no doubting that the specter of monopoly that the ABC raised and the concerns over corruption justified by its (and other publishers') employees led to state-level regulation of the industry. Though this was not intentionally about textbook content, the industry's organization of its curricula, particularly in biology, utterly failed to address the realities of textbook adoption in states like Tennessee. Out of this neglect, the possibility of the Scopes trial became real.

CHAPTER THREE

Textbooks and Their Makers: Authors, Editors, Salesmen, and Readers

In December 1923, Charles J. St. John, a sales agent working for the American Book Company (ABC), attended Mississippi's annual High School Conference. He wrote to W. T. H. Howe to relate "what thoughts are coursing through the minds of [his] constituents." The letter included assessments of the ABC's prospects in literature, math, science, and history. "I wanted to see a recommendation of a course combining Ancient and Modern History into one year. I had our Elson's 'Modern Times and Living Past' in mind," St. John wrote. "I had two friends advocate that course on the floor but it seems that a great majority prefer Ancient in the first year, Modern in the Second. . . . This too was referred to committee for report one year later and I shall work to sell my idea to the committee."[1] St. John's report to Howe gives a candid picture of textbook salesmanship in this era, complete with accounts of some practices that state regulation was intended to curtail. It also shows the rationale behind some of these practices. "Friends" were used, not only to influence adoptions, but also to alter curricula, a system made all the more necessary by a basic truth of textbook selling at this time: it was often much easier to adapt a state's requirements than to inveigle publishers to change the textbook.

St. John's letter to Howe also reveals how complex textbook production was in this era. Textbooks were not simple creations—conceived by an author, revised by editors, and given over to salesmen to turn into profit. The process of making a textbook was not so unidirectional. Sales agents were also the primary source of information about a book's audience and readers' expectations. Letters like St. John's were routinely forwarded or summarized by managers like Howe in communications to editors. Sometimes, they were even passed back to authors.

In many parts of the country, sales agents had nearly total control of the flow of information between publishers and textbook users. This enabled them to influence managers and editors. More importantly, it also made them the local authorities on the textbooks in the places they worked. Even when sales agents could not lobby their publishers to change something in a book or convince a state or school board to alter its curriculum, they still had a great ability to guide the way consumers read and interpreted books. Even though St. John did not manage to sell the ABC's history book at the Mississippi High School Conference, he had another year to "sell [his] idea" before the next history book adoption.

Selling a book included convincing regulators that a text taught what schools said they needed. Especially in places where adoptions were mostly determined by judging a textbook's physical characteristics—price, availability, and durability—the people selecting textbooks may not have had either the time or the expertise to evaluate all books submitted in every subject. A salesman with a long-standing relationship in a community was often seen as credible, and his authority was usually bolstered by testimonials, favorable book reviews, endorsements by experts, and lists of other districts and states that had already adopted the book. Sales agents often traveled with copies of these materials, using them to suggest how each textbook should be used. Even when textbook salesmen were not taken as trustworthy allies of the school, in places more wary of conflicts of interest, they still had "friends"—teachers, other school officials, or pedagogy experts—whose credibility had currency and whose self-interest was less overt, to help them out with suggestions "on the floor."

The influence of these salesmen was strongest when it was assumed (both by salesmen and by school officials) that teachers would not diverge from the textbook in the classroom. The farther one went from the cities where most textbooks were published in the United States, the farther one was from the centers of widespread public education and high concentrations of teachers who had any special training or the professionalizing effects of local pedagogy experts and teachers' unions. In these regions, teachers were expected to follow a textbook. Governor Austin Peay revealed this expectation when he signed the antievolution bill into law, stating: "I can find nothing of consequence in the books now being taught in our schools with which this bill will interfere in the slightest manner. Therefore, it will not put our teachers in any jeopardy."[2] Tacitly, Peay reinforced the impression: as it was in the books, so it would be in the classroom. Although education experts of the era sometimes fretted over practices of the rote learning that textbook-based pedagogy encouraged, they acknowledged that the textbook could be a source of training for teachers who had little background in pedagogy. Textbooks were expected to provide an intellectual rigor in places where untrained teachers could not. Herman L. Donovan, the dean of Eastern Kentucky State Teachers College and an alumnus of both Teachers College in New York and the Peabody School of Education in Nashville, summed up the situation in a 1924 article about textbook selection: "The poorer the teacher, the better the textbooks need to be."[3]

Students, teachers, and state officials all regarded textbooks as valuable teaching tools precisely because they were taken to be authoritative. The credibility that sales agents had as mediators between publishers and consumers became transferred to the textbooks they spoke for. Add to this the imprimatur of state adoption, and a situation emerges where textbooks were regarded by readers as a verified, authorized, and legitimate source of prescribed knowledge. Textbooks were also validated by the reputations of the authors and publishers whose knowledge and training were expressed in a textbook's creation. As the education researcher David Olsen would describe it decades later, textbooks were taken to be a "transcendental source" of knowledge, concealing any marks of fallible human creation.[4]

But, obviously, textbooks have human creators. Who are they? In the history of publishing, printing, and reading, authorship has long been a contested notion. In the early history of printing, attributions of authorship were often obscure, misleading, and met with skepticism by readers.[5] Readers in the twentieth century tended to regard authorship as more stable (though many of these issues have returned in the postprint digital age). Nonetheless, textbooks were often identified with an author, so much so that biology books were often referred to by the names of their authors, for example, *Hunter*, *Moon*, and *Peabody and Hunt*. Sometimes, the publisher's name also had a similarly strong association with a textbook or series. Publishers often used authors' names as a type of brand whose reputation could sell books.

In addition to authors and editors, salesmen and state regulators shaped the authorship process. Communities of readers were anticipated and their expectations and habits perceived and transmitted to others involved in textbook production. To the extent that sales agents were in a position to influence the content of textbooks and school standards and the way in which these two things related to one another, they played a tremendous role in shaping the perceived authority of textbooks. These perceptions of authority shaped the way in which audiences read and interpreted the texts. If the creation of a textbook is taken to include the intangible efforts that go into making a book into a textbook, sales agents were as much creators as anyone else in this process. More than selling bound volumes of printed paper, men like Charles St. John were really selling ideas about how to read the books in their catalogs.

Perhaps the most important idea that salesmen sold was that of their own invisibility. Even though the names of authors and publishers stood as guarantors of quality, an effort was made to present the textbook as an object without a history—without a human creator. This was especially true of science textbooks, in which notions of objectivity were valorized. "Objectivity," as Lorraine Daston and Peter Galison put it, "has a history" and developed as a virtue in a specific context. In Daston and Galison's historical analysis of the subject, the early twentieth century was the time when notions of an apersonal, "noninterventionist" view of the objective ideal were beginning to be supplanted by respect for the judgment of the trained expert.[6]

While the chronology of the virtue of objectivity in textbooks may not coincide exactly with that of scientific practices, it appears that notions of both mechanical objectivity and trained expertise were central to educational practices and the role of textbooks in this period. The names and reputations of individual authors were important as a mode of branding and assuring the qualities of a textbook, but those qualities that were being assured were those of objective science. Patterned after "knowledge that bears no trace of the knower," the pedagogical ideal of objectivity bears no trace of the teacher.[7] Untrained teachers would not contradict material given in the book. They would use the teaching manual and would lack sufficiently trained judgment to innovate beyond the activities, lessons, and sample questions given by a text.[8] In this way, they would mechanically transmit textbook contents to students. An education system that relied on this kind of teaching illustrates how pedagogical objectivity was expected and made normative. This was by no means exclusive to science teaching, but the role of objectivity in scientific practice makes the connections in the scientific disciplines especially notable.

The view of textbooks as objects with little to no human character has figured prominently in the way in which historians of science have used them. This view is, of course, incomplete. It presupposes that textbooks serve a very authority-heavy audience—used either to dictate the truths of science to the unscientific populace or to be read by people being trained to take up roles as scientists. In either case, it is assumed that readers *must* accept textbooks as authoritative.[9]

Stripping away the mantle of invisibility that textbook sellers use to cloak their products, textbooks are revealed as much more complex objects—even in science. As a human production, a textbook is not simply an epitome of a scientific field. Textbooks bear the marks of the individuals who write and edit them. They also develop out of the cultural norms of the communities that create, distribute, and read them. At the primary and secondary levels, education is also regulated by a variety of authorities who have other goals than the training of scientists. Consequently, the creation of texts for these education levels is even more complex.

The history of high school biology textbooks in the early twentieth-century United States provides a striking example of how scientific knowledge has been produced and distributed to nonspecialists. To some extent, these textbooks do represent the various developments in biological knowledge, but they were also deeply shaped by the structure of textbook production and distribution. Rather than viewing high school biology textbooks as the products of a single integrated community, one ought to recognize the separate efforts of two very different communities: one of biology authors, small, collaborative, and ideologically committed to a scientific worldview; and another of textbook salesmen, diffuse, competitive, and dependent on maintaining a network of personal contacts for their livelihood. In many ways, the two communities could not have been more different. Their social structure, their politics, and their views of the purposes of textbooks contrast sharply. Occasionally, individuals from the one community interacted with individuals from the other, but most of their communication was indirect, mediated by textbook editors. These two communities were both crucial to the formation of the modern high school biology textbook, the object of scientific knowledge ultimately placed into the hands of teachers and schoolchildren around the country. Despite the presumptions of teacher and student reading as *following* (as passive rote learning was often derided), the classroom was a site where interpretation and use of the textbook could vary. The variation in how a text was used was in no small part due to the almost contradictory aims of salesmen and authors. The only thing that seems to unite them is a common interest in the production and distribution of biology textbooks. But, when it comes to why and how such books should be created, these two groups could not have felt more differently.

A COMMUNITY OF RIVALS—TEXTBOOK SALESMEN IN THE EARLY DAYS OF BIOLOGY

Even though Charles St. John mentioned Mississippi's concern for science and history curricula, his December 1923 letter to W. T. H. Howe was not principally about the content of his books. It was about his competition with other sales agents and especially the ABC's ongoing rivalry with Ginn. The bulk of the letter makes this plain:

> Mr. Newell of Ginn & Co. was in the conference once. He was not received very cordially and did not return. I lost no opportunity to advertise his dirty work of the summer months.
>
> Mr. [Henry L.] Louis had left it to my discretion as regards leaving this territory. I think now that I will get back into Miss. the last of this week. Work here is not satisfactory now. The schools, teachers as well as students, are full of Christmas spirit and I feel it in my efforts.[10]

In the persons of St. John and Newell, the ABC-Ginn corporate rivalry, dating back to 1890, took on human form. Sales agency was the primary way in which employees of one publishing company were likely to come into direct contact with those of another. Individual sales fights across the United States helped forge the rivalry between the companies. The ABC and Ginn engaged each other on the corporate level as well, publishing pamphlets denouncing one another, and raising counteraccusations of slander or monopolistic practices.[11] These denunciatory pamphlets were key ammunition in the sales agent's arsenal, given to prospective customers just as readily as brochures touting one's own wares. It was just as important for St. John to "advertise [Newell's] dirty work" and attack Ginn as it was for him to advocate for the ABC's books. Men like St. John depended on their ability to sell to earn their livelihood.

Not only were sales agents one of the primary points of contact between rival publishers, but they were also the principal interface between publishing companies and the public, including the people who bought and used textbooks. Within the structure of the ABC, information about a region was collected by men like St. John and passed to local managers like Henry L. Louis, in Nashville, or to managing directors like W. T. H. Howe in Cincinnati. Only then was it transferred to the editorial departments in New York, from Howe to George W. Benton (the editor in chief) or W. W. (William Winfred) Livengood (the assistant editor in chief). Editors then contacted the printing department in New Jersey to order more copies of books or communicated with textbook authors to pass along suggestions, criticisms, and feedback.

The ABC had a more complicated hierarchy than most publishers. This was partly an artifact of the several companies whose merger formed the ABC. In addition to its headquarters in New York, it had major offices in Boston, Chicago, and Cincinnati. Most other companies were not large enough to have major regional offices and instead had individual sales agents report to local sales managers, who reported directly to a company's main office (where their editorial and printing offices were also based). Even in smaller companies, this organizational model ensured that sales agents were not just the primary point of contact between different companies but also the publishers' direct contact with the general public. Editors and textbook authors were not expected to travel and, thus, relied on sales agents to obtain reviews and reports about how their products were received in far-off classrooms. Sales agents were the face of their respective companies, and the publishers relied on the eyes and ears they had in the field. At the same time, their books' reputations often relied on how their consumers perceived that peculiar physiognomy.

THE TEXTBOOK SALESMAN IN AN AGE OF REGULATION

Although the regulation of textbooks at the state level would not necessarily have hurt textbook companies, it did threaten the structure of sales agency since fewer (albeit larger) adoptions required fewer agents. Sales agents could not fail to report on the new adopting environment, but their information led many companies, especially the ABC, to lobby against statewide regulation, rather than to instruct the editorial offices to adapt content accordingly. Not only was the prospect of state regulation threatening the livelihood of textbook sales agents by creating an environment where fewer were needed; it also made individual adoptions more difficult and, with so much at stake, more stressful for the agents lobbying a state. As Howe told Editor in Chief George Benton in 1923: "You have got to keep the morale of your agency force to white heat in these state adoptions."[12] Sales agents worked long and tiring hours, building personal networks, and advocating for books on a variety of subjects. Given these pressures, it really is unsurprising that the history of textbook regulation is rife with instances of salesmen resorting to illegal methods.

The changes in textbook regulation and the evolving climate of textbook marketing led to concern at Ginn over the morale and vigor of the company's sales force. Noting that its sales had fallen in 1925, Ginn's sales agents considered the crisis of their role within the company: "Had we not brought out a single new book, should not our sales have increased on account of the growth in population? . . . One thing seems certain: either the sort of books which we have brought out on the market are not what the people want, or else we have been deficient in our methods of putting them on the market." The real issue in 1925 was whether textbook markets had so changed as to render sales agency nearly obsolete. The memorandum of the agents' meeting was candid, if inconclusive:

> Is this falling off due to the fact that the members of the firm have lost some of their snap and keen interest in the business and that this lack of energy and get-up-and-git is reflected in our agents; or is it due to the fact that our agents have been with us a long time and have lost a lot of their energy and aggressiveness? Is it that the use of textbooks is dependent more and more upon the superintendent and teachers and less upon what the agents can say; or is it due to the fact that a new kind of agent on educational matters with something of a missionary spirit has come out of the field and that good fellowship is less needed in this business than formerly? In other words, is the personal element likely to be as strong a factor in the future as it has been in the past? Or is it due to the fact that our agents lack defensive ability and always want something new?[13]

Textbook salesmen themselves had become aware of the changes in their profession. As textbook regulation grew and the tactics of an older generation gave way to newer ideas of salesmanship, "the personal element" seemed to become a relic of a bygone era. Ginn president George A. Plimpton had also taken note of the evolving climate of textbook sales: "I think we are to be congratulated on our agency force. They are men who inspire confidence in superintendents and teachers, and their sales, with few exceptions, justify their salaries." Nonetheless, Plimpton tempered this praise with deep concerns for the future of these methods: "The disposition of many of our men after they have been in the business fifteen or twenty years is simply to touch the high points but not to go after the little fellows who, sooner or later, will supplant the big men. Now where we have cases of this kind we should have a young man to put in as a sort of assistant to take the place of the old man or to work with him. . . . There is an army of young teachers growing up who are not meeting our agents."[14]

With increased state regulation, a larger voice for pedagogy experts, and the greater professionalization and training of teachers, the heroic (or villainous) sales agents of the nineteenth century had given way to a generation whose tactics seemed to be obsolete by the mid-1920s. Whether their failings were the result of the new marketing environment or a reflection of the quality of the men was a matter for debate. But textbook companies had not taken steps to replace sales agency with a system better suited to more regulation. In 1926, Plimpton was only beginning to question the effectiveness of sales agency for his company. Edwin Ginn, the company president until his death in 1914, had begun his career as a textbook salesman. Plimpton had worked as an editor at Ginn, and perhaps this was why he could be more openly critical of sales agency itself. But he also had to acknowledge his company's dependence on sales agency, a dependence that the salesmen of Edwin Ginn's generation had cultivated. As sales agency would fall apart without competition, the community was held together by its own codependence. If any company failed to compete or became too successful, the future of sales agency as a whole would suffer. Whether members of the sales forces of these companies acknowledged it or not, they had more in common with one another than with the editors or corporate executives who worked for the same company and perhaps least in common with the textbook authors, who were not directly employed by these textbook companies but provided the raw materials that became their wares.

This then was the culture of textbook salesmen in the late nineteenth and the early twentieth centuries. They constituted a community driven by their occupation; their salaries were justified by the sales they generated. To survive, they established close personal contacts with teachers, school superintendents, and school board members. At times, these close contacts led to ethically or legally questionable conduct. Bribery, kickbacks, electoral interference, and collusion were acceptable in dealings with school officials, as were slanders about and other attacks on rival agents. Their assignments to geographic regions and the hierarchies through which they were isolated from the editorial departments of their companies led them, in many cases, to form closer ties with their customers than with employees from other divisions of their companies. They were expected to operate independently and with little oversight. For the most part, they were not expected to focus on textbook content. Yet they were also expected to serve as conduits of information for editorial departments. Toward the end of this era, their influence with consumers waned. Growing numbers of teachers trained in the new *science* of pedagogy and more influenced by education experts led to the erosion of their personal relationships. So too did increased government regulation. Yet, even into the 1920s, salesmen wielded substantial power over publishers themselves and, through this power, shaped content and policy.

As competition intensified, many younger textbook salesmen were salesmen first and textbook men second. In this respect, educational publishing was no different from many other industries that began to make greater use of the salesman in the early twentieth century. In fact, the sales agency model for textbooks preceded the rise of similar salesmanship strategies in many other U.S. industries, perhaps influencing the culture of salesmanship that emerged in the late nineteenth century. In turn, the culture of salesmanship may have reduced the clout of textbook salesmen in the 1920s.

SCIENCE, SALESMEN, AND SALESMANSHIP

The rise of textbook sales agency in post–Civil War America coincided with the increase in national textbook markets. The expansion of railroads, the development of less expensive printing technologies, and the growth of public education in the late nineteenth and early twentieth centuries facilitated the creation of new publishing companies and led to more geographic overlap between the markets each pursued. This led to more direct competition.

The rise of textbook sales agency also coincided with the formation of a larger culture of salesmanship that saw itself as scientific. In the late nineteenth and early twentieth centuries, many industries in the United States began to hire people to work specifically as salesmen. The textbook industry was among the first, but other publishers also "organized large teams of canvassers to sell popular books." As Walter A. Friedman observes in his history of American selling, booksellers quickly became an object of resentment.[15] The textbook salesmen were similar, though their potential clients were more specific to school officials. Though this meant that they were less ubiquitous than, for example, encyclopedia salesmen, the fact that their clients were government bodies prompted greater worry about corruption.

Ironically enough, the rise of the new commercial practice of salesmanship was testified to and facilitated by new textbooks on salesmanship. Thomas Russell's 1910 *Salesmanship Theory and Practice*, one of the earliest salesmanship textbooks published, proclaimed the novelty of the practice: "Up to within the last few years the word 'salesmanship' was not to be found in the dictionaries of the English language. It was not recognized as a distinct word."[16] The *Oxford English Dictionary* traces the earliest occasions of *salesmanship* to the 1880s, when it referred to a quality of an individual; certain people have a natural talent for salesmanship. The *OED* dates the sense of *salesmanship* as a *practice* to the 1930s, but these textbooks ushered in this use decades earlier. Salesmanship textbooks also took note of what they considered to be the sudden interest in the practice. A 1925 textbook offered the statistic that there were only 5 books listed in the Library of Congress catalog under the subject of salesmanship published through 1890. From 1890 to 1900, another 6 books on the subject were published. From 1900 to 1910, 36 books on salesmanship appeared and, between 1910 and 1920, 220 books.[17]

These salesmanship textbooks emphasized certain techniques for understanding one's customers, using psychology, and understanding negotiating. Frequently, these books referred to salesmanship as a *science* and particularly as one rooted in the principles of evolution. Russell proclaimed: "The salesmanship of today is often called scientific salesmanship."[18] The first chapter of a 1920 YMCA textbook was entitled "The Evolution and Scope of Business."[19] The idea that salesmanship was an outgrowth of an evolutionary principle in industry was frequently made explicit. The very existence of a community of salesmen, Russell argued in a 1924 textbook, was best understood in terms of social evolution: "Following the universal rule of evolution, business control has developed from the simple to the complex and varied. Functions not long performed by the same individual have differentiated, and have been distributed to several persons."[20] Thus, salesmen who came to practice in an era of scientific salesmanship had already imbibed the authority that was involved through the labeling of the practice as *scientific*.

The discussion of the differentiation and distribution of business functions as an evolutionary process might suggest a cultural bias in favor of science on the part of the salesman. But, if evolution and science were essential principles to the doctrines of salesmanship in other industries, it hardly mattered to most textbook salesmen—even when the textbooks they were selling taught evolution and other principles of science. (This would be seen shortly after the Scopes trial as many agents were very quick to encourage editors to remove evolution from biology textbooks.) American textbook sales agency preceded the rise of salesmanship outlined in these textbooks of the first two decades of the twentieth century. These textbook salesmen did not read salesmanship textbooks; they sold them. Their practices had developed in the late nineteenth century, before scientific salesmanship was conceived.

The generational issue that Ginn and Company began to recognize in the 1920s—the loss of the personal element and a population of younger teachers who had no direct relationship with their sales agents—shows the effects of a transition to a new culture of salesmanship. The difficulties that textbook salesmen faced in the 1920s stemmed, not only from increasing regulatory pressures, but also from the fact that their tactics, honed from the 1870s through the 1890s, had been surpassed in other industries.[21] As scientific salesmanship became more common in other industries, the expectations customers had of textbook salesmen also changed, and relied-on tactics became untenable. Regulatory practices may have exacerbated this and contributed to the woes of textbook salesmen in the 1920s.

The creation of salesmanship as a practice also meant that people were trained in a skill that did not depend on the particular items being sold. The very idea that one could learn from a textbook on salesmanship implied that it did not matter *what* one was selling. Salesmanship textbooks were possible because selling was not product specific. The personal instincts of the sales agent and the rapport developed with customers were most essential. The quality of a product was not completely irrelevant, but a salesman was taught to sell the customer, not the product. This was consistent with the approach of textbook salesmen, which involved not emphasizing the specific content of individual books. In this respect, scientific salesmanship resembled the practices of textbook salesmen.

Sales agency was separated both physically and occupationally from the editorial and executive departments of textbook publishers. However, there were occasions, such as in the transmission of feedback, when roles intersected. Many of the founders and executives of textbook companies began as salesmen. Consequently, the salesman mentality influenced other parts of a company's practices. Many of the companies were quite profitable, yet they cultivated the perception that they were on the brink of bankruptcy; maintaining and increasing sales was vital. This mind-set, along with the feeling that publishers kept prices artificially high, led advocates of textbook reform to question whether private, for-profit textbook publishers actually served the public interest. Could companies be trusted to provide the best materials when they were more concerned with whether a book could sell than with whether it could teach?

Edwin Ginn, salesman-turned-publisher, invoked this rhetoric himself in pamphlets challenging the ABC. By doing so, he tried to project an image of the textbook publisher as a responsible citizen. This was reinforced by his very public philanthropy, although with mixed success.

PROFIT, PHILANTHROPY, AND SELLING TEXTBOOKS

Claims of education quality were bound up (in some cases literally) with claims of altruism or public service. In the pamphlets that Ginn and the

ABC issued against one another in the early days of their rivalry, each side accused the other of being driven by profit while claiming that textbook publishing ought to be a profession with a nobler purpose. Profit and public service were both motivations for many people in this business. Publishers were in the business to make money, yet many of them had come into the particular industry of textbook publishing out of some concern for promoting education.

Edwin Ginn's philanthropy demonstrated how entangled these motivations could be. Ginn became deeply involved in the movement for international peace in the 1890s.[22] "We spend hundreds of millions a year for war," he told the 1901 meeting of the Mohonk Conference on International Arbitration. "Can we afford to spend one million for peace?"[23] In 1910, he helped found the World Peace Foundation and, on his death in 1914, bequeathed to it that million dollars.

In addition to donating money, Ginn also made use of his publishing kingdom to advance this cause. As early as 1905, Ginn and Company published materials "designed to serve as textbooks at a peace school."[24] Like that of his fellow businessman and international peace advocate Andrew Carnegie, Ginn's commitment to world peace was likely rooted in a belief that peace was better for business. Arthur N. Holcombe suggests that Ginn was also inspired by "the theory of social evolution, expounded by Herbert Spencer in his *Synthetic Philosophy*."[25] This application of evolutionary science to the development of human societies was connected to the biological and physiological theories of evolution but tended to emphasize the ability of human societies to advance and direct their own progress. Spencer's theories of social evolution supported a view of a human species that was progressing beyond its primitive, warlike nature toward a peaceful civilization. Spencer's ideas were especially influential among some educational psychologists of the era, and his books were sometimes taught in schools of education. Ginn had no formal training in pedagogy, but he was certainly well versed in its trends. Spencer was also very close to Andrew Carnegie, who described himself as one of Spencer's "disciples" in his autobiography.[26]

Ironically, these social evolutionary peace advocates had common cause with the antievolutionist and pacifist William Jennings Bryan, who also addressed the Mohonk conferences, and some of whose speeches there were published by Ginn's World Peace Foundation.[27] Some historians have attributed Bryan's antievolutionism in the 1920s to his horror of the militaristic uses of social evolution and eugenics in pursuit of the Great War.[28] Yet the fact that Bryan and Ginn shared the same goal of international peace complicates the idea that evolutionary thought led to the war. This common cause was also the basis of a relationship between Bryan and Ginn's company that continued even after Edwin Ginn's death in 1914.

Ginn's philanthropy and his use of his publishing house to promote his social agenda speak to *something* having priority over profit. International arbitration and world peace as it was understood were not popular in the United States, especially in the first years after the Spanish American War. Ginn and Company's reputation was not hurt by its founder's political views, but his business partners were concerned that it might be damaged by his philanthropy. They feared that Edwin Ginn's use of his fortune would reveal to the public just how profitable textbooks were. Noting that textbook publishers were frequently criticized for "the high price of school books in this country" and "the huge profits they are making in this business," one of the partners in Ginn and Company took exception to Ginn's promise of a million dollars for peace: "Mr. Ginn has a perfect right to use his money for whatever purpose he pleases, and the more good he can do with it, the better for the world. I don't believe that he has the right, however, to publish a benefaction of this kind . . . to the embarrassment of his partners and the business. What we ought to do is make every effort to disabuse the public of the notion that this business is largely profit, and not give them the impression . . . that it is a source of great wealth to those who are engaged in it."[29] Strong salesmanship was important, not only in selling textbooks themselves, but also in promoting a good reputation for the textbook industry and those who profited from it.

THE SALESMAN TURNED PUBLISHER

Edwin Ginn got his start as a publisher by purchasing the plates of one of the books he had previously sold as a salesman.[30] A year after he died, one of his company's salesmen did the same thing. Edward F. Southworth founded the Iroquois Publishing Company in Syracuse. On July 15, 1915, Ginn and Company issued a letter to its sales agents:

> We have sold to E. F. Southworth the See and Say Series. These books will be handled in the future by the Iroquois Publishing Company, Inc. of Syracuse, New York, which has been organized by Mr. Southworth. This sale has been made at Mr. Southworth's request, so that he may be free to devote his energies to pushing books in which he is personally interested.

> This termination of a relationship which has existed between Mr. Southworth and our firm for many years has been made with the friendliest of feelings on the part of all concerned.[31]

From Southworth's acquisition of a single series of basal readers (which Ginn still distributed), Iroquois grew.[32] Though it was never one of the larger publishers, it eventually opened regional offices in New York, Chicago, Atlanta, and Dallas in addition to its headquarters in Syracuse. "In less than a quarter of a century," a 1939 advertising poster stated, "Iroquois has assumed nationwide proportions."[33]

Southworth had a personal connection with the *See and Say Series* that was unusual for a sales agent. He was named as the books' third coauthor.[34] Over the next few years, he added a few titles to Iroquois's catalog. Many of these were not fully textbooks but study manuals, written to prepare students for state examinations in New York. The first science books that Iroquois published were of this nature. In 1922, it brought out the *Laboratory Guide for Chemistry* and the *Laboratory Guide for Biology*.[35] Both of these were lists of experiments and study questions meant to prepare students for the New York State Regents Examination. The readers and some of the other early books could, however, be used elsewhere. From 1915 to the early 1920s, Iroquois was largely a regional publisher.

While most of Ginn's agents worked a single territory over the course of their careers, Southworth had canvassed several areas, including New England, New York State, and Washington, DC.[36] This experience and the contacts he maintained helped him expand Iroquois's presence. he remained personally involved in the sales operations of his fledgling company and, especially in the early years, traveled extensively to secure sales.

In June 1925, when Tennessee considered textbook adoptions (after the 1924 postponement and just a month before the Scopes trial), Southworth was in Nashville, advocating for his company's books. He introduced seven books for consideration, including the *Laboratory Guide for Biology* and the first science textbook published by the company, Arthur Clement's *Living Things: An Elementary Biology*.[37] Contrasting Clement's book with the ABC's *Civic Biology* by George W. Hunter and Allyn and Bacon's *Practical Biology* by Smallwood, Reveley, and Bailey (the two books adopted in 1919 and still in use), Southworth noted that *Living Things* was less expensive than these competitors. He continued: "Living Things is less than a year old and yet it has already been adopted by over four hundred systems of size; an unusual book peculiarly fitted for state wide use."[38]

Emphasizing how new a textbook was had become a standard practice for textbook salesmen, but in this case it was especially poignant as Tennessee had enacted the antievolution law just two months earlier. Southworth's insistence that Clement's book was "peculiarly fitted for state wide use" was meant to imply that it was compliant with the new law. The book at the center of the Scopes trial, Hunter's *Civic Biology*, had really been intended for use in urban schools, not across a state that was largely rural. *Living Things* had been written with a single state's—New York's—standards in mind. But Clement did not have to do much to adapt the book to New York's standards for the very simple reason that he had largely written those standards himself.

THE COMMUNITY OF BIOLOGY AUTHORS—A RESPONSE TO INTRUSION

When *Living Things* was first published, there was an immediate outcry. It did not come from textbook salesmen, antievolutionists, or textbook regulators. Instead, Clement's book outraged other biology textbook authors. They were not upset by the book's treatment of any biological subject or Iroquois's sales tactics. Instead, they were furious because Arthur Clement had been supervisor of biological sciences in the New York State Department of Education until just before the book's publication. His book was ideally suited to the revised New York State standards that had come out earlier in 1924. In an advertising circular sent to biology teachers and department heads in New York in May, Southworth bragged: "Living things—An Elementary Biology—By Arthur G. Clement is built on this new Syllabus. It follows in every detail the Syllabus order. The book is divided into five parts corresponding to the Syllabus divisions."[39] The advertisement did not mention that the book followed the syllabus so well because the standards had largely been written by Clement.

Benjamin C. Gruenberg, then the Biology Department head at Julia Richmond High School for Girls in Brooklyn, was one of many teachers to receive this advertisement. He was also the author of the 1919 *Elementary Biology* and was working on a new high school biology textbook that would be published by Ginn in 1925. In fact, most of the men and women who wrote high school biology textbooks in the 1910s and 1920s were faculty in schools or universities in New York State. Nearly all of them received a copy of Iroquois's circular. This was the first any of them had heard about Clement writing a textbook. Suddenly, they realized why they had had so much difficulty in acquiring copies of the new standards that many of them had requested in order to write and revise their books.

Trickery was usual for textbook salesmen, but among biology textbook authors cooperation was the norm. Clement's behavior was well beyond what was acceptable. The fact that his book gave rise to a unified reaction from the authors of nearly every other major textbook shows how much biology textbook authors valued that collaborative spirit. They were not employees of the rival companies who published their books. It seems that they felt greater loyalty to one another than to their editors or their companies' sales forces. Though they disagreed with one another on details of pedagogy and their books reflected differences of approach and scope, they did regard each other as collaborators, working together in the cause of a more engaging and robust education in science. This community among authors justifies speaking of the new biology curriculum of the 1910s and 1920s as a coherent movement that goes beyond the commonalities of the books they wrote.

When Clement's book came to the attention of these biology textbook authors, they immediately cried foul. "I wrote to the State Department for a copy of the new proposed syllabus in biology. Mr. Clement replied to the letter, and in sending the syllabus wrote that he would be glad to have suggestions for any needed changes," Henry Linville wrote to Gruenberg. Both men were New York City biology teachers, authors of high school textbooks, and among those most responsible for the development of the biology curriculum of the first two decades of the twentieth century. Linville also observed: "Being in an official position that gives him considerable power in deciding the content of the syllabus, the natural tendency would be for the syllabus to be adjustable to Mr. Clement's book, while other textbook[s] in the subject would have to be adjusted to the syllabus."[40] Textbook salesmen were already in the practice of lobbying for changes to state standards through their "friends." Clement was not a salesman, but, as Linville observed, he was in a position to do the same thing even more directly. Such a tactic was unavailable to other textbook authors.

Gruenberg had been inquiring about the revised state syllabus for over a year when Clement's book came out. For that whole time, he was unable to obtain a copy, despite the assurances of other Department of Education officials that "we intend to give every publisher and every author exactly the same chance."[41] But every author did not have the same chance, and many of them wrote back to Gruenberg voicing displeasure. "I certainly appreciate the fact that you have finally discovered why we were unable to get access to the new Syllabus until late in December," William M. Smallwood, based in Syracuse, wrote to Gruenberg.[42] George W. Hunter, the author of *Civic Biology* and a former New York City biology teacher, noted: "Clement is going to retire in a short time. This may be a reason for his writing a book."[43] For this reason, he suggested, Clement had no need to worry about burning bridges with his government position. The book so enraged James E. Peabody, a biology teacher at Morris High School in the Bronx, that he wrote Gruenberg: "I wish I was not an author of a biology just at the present time and I would take off my coat and try to make things fly."[44]

This was not the first time that something like this had happened in New York State. A year earlier, Gruenberg mentioned to another colleague, Otis Caldwell, that "a committee appointed to prepare a special syllabus dragged its work out for about two years; and when the syllabus was finally published officially, a textbook on the subject, by the chairman of the committee, appeared simultaneously."[45] Despite Gruenberg's hope that the state commissioner of education would not allow a repetition of such an abuse, Clement's textbook was exactly that.

Gruenberg spearheaded a series of circular letters among many of the biology textbook authors discussing what response to make. Gruenberg's letters were addressed to nearly every biology textbook author of the 1910s and 1920s: Ida L. Reveley, Otis W. Caldwell, Maurice A. Bigelow, H. A. Kelly, Henry R. Linville, George W. Hunter, James E. Peabody, Arthur E. Hunt, William M. Smallwood, Guy H. Bailey, William H. Atwood, Gilbert H. Trafton, and Truman J. Moon.[46] There was a strong consensus that some sort of group action ought to be taken. Realizing that they were powerless to actually change Clement's textbook or the use of it in various parts of the state, many of them cosigned a letter of protest to the New York commissioner of education. Some authors decided not to sign because the letter could not lead to any real change. "If this protest is forwarded to Commissioner Graves," wrote Peabody, "he will rightly demand names and dates. Are any of us prepared to give them?"[47] Despite Peabody and his coauthor, Arthur Hunt, deciding not to sign the letter, there was still a sense of unity that shows how these authors regarded one another as part of a close-knit community.

The camaraderie between biology authors is striking given the fierce competition between their publishers. Peabody and Gruenberg were old friends; when Peabody and Hunt were about to publish two textbooks in 1912 (one on plant biology, the other on animal and human biology), Peabody asked Macmillan to send Gruenberg the manuscripts. "Rip it up, please!"[48] he implored his friend. Gruenberg was not yet writing for Ginn, but it was not uncommon for these authors to discuss their upcoming work and even to collaborate, even when working with rival publishers. In the 1910s, Frank M. Wheat prepared illustrations for both Hunter's (ABC) and Gruenberg's (Ginn) textbooks. Wheat later became an editor at the ABC and in 1929 coauthored a biology textbook for upper-level high school and introductory college levels. The 1929 *Advanced Biology* by Wheat and Elizabeth T. Fitzpatrick included a dedication "to two successful pioneers in biology textbook making, George W. Hunter and James E. Peabody, from whom the authors of this book received their early training and inspiration."[49] Hunter and Peabody never collaborated directly, but they were "pioneers" on the same frontier. The fact that their books were competitors (a book coauthored by Peabody replaced Hunter's *Civic Biology* in Tennessee just before the start of the Scopes trial) did not signify. Wheat and Fitzpatrick had corresponded long before their own collaboration, discussing teaching techniques using the Wheat-illustrated *Elementary Biology* by Gruenberg.[50]

This community, and their collaborations, shaped the new pedagogy of biology in the 1910s and 1920s. The intimacy and influence of this community meant that modern high school biology was formed under a shared set of norms that helps explain why the scientific subject of biology came to be seen as promoting specific cultural values. The coherence of this community was also aided by geography. The proximity of these teachers and authors to one another allowed for greater collaboration and interaction, and their location in an urban, industrial, and culturally diverse state helped shape the values they jointly espoused.

Several textbook authors even taught at the same schools. As Philip J. Pauly has described, the epicenter of this was DeWitt Clinton High School in New York City: "Clinton's biology department, which included thirteen men by 1912, was led by three individuals at the forefront of the transformation of American education in the early twentieth century."[51] These were Linville, Gruenberg, and Hunter. In addition to these men, Frank Wheat and William Atwood, as well as Max Mandl and Nelson S. Kline (who also wrote biology textbooks), spent part of their careers teaching at DeWitt Clinton. Many other authors were also based in New York City or nearby; Maurice Bigelow was at Teachers College, and Peabody and Hunt taught in New York City high schools. Even those outside the metropolitan area were close enough that rapid communication (with one another and with publishers) was possible. Truman Moon was in Middletown, New York, and Smallwood, Reveley, and Bailey were in Syracuse. As a result, the New

York State standards held especial importance for these authors, which made the circumstances surrounding Clement's textbook all the more important to them.

The community of biology textbook authors was collaborative, geographically close, and less connected to publishers than to one another. In all these respects, it was very much unlike the community of textbook salesmen. These two communities also differed completely on what they saw as important about the high school biology textbooks that they were all involved in creating and distributing. For the authors, unlike the salesmen, nothing mattered more than the books' content. The two parties' views on the whole process of textbook production were inimical. They often clashed, creating difficulties for the people who mediated their relationship: textbook editors.

TEXTBOOKS, KNOWLEDGE PRODUCTION, AND THE CONNECTIONS BETWEEN COMMUNITIES

In 1923, W. T. H. Howe complained to ABC editor in chief George W. Benton about the ABC author Charles T. McFarlane. As the controller of Teachers College, McFarlane had just hired the superintendent of Fort Smith, Arkansas, C. J. Tidwell. According to Howe, Tidwell was "ridiculously antagonistic" to the ABC, even telling one of Howe's agents: "'So far as I am concerned, I would bar the American Book Company from the state for ten years just for a reminder of their interference in Arkansas last summer in the election at various times in the past.' [Tidwell] also said to this same agent that he told Dr. McFarlane that it was unfortunate indeed that the American Book Company had the publication of his geography." Incidents like this, Howe claimed, made it difficult for sales agents to do their work: "They like to push the Brigham and McFarlane Geographies because they are good geographies, but when a fellow who is out fighting for his life in the state adoption runs up against a thing like this there is no possible means for him to feel anything but opposed to Dr. McFarlane or anything his name is associated with."[52] Howe's portrayal of his sales agent as a "fellow who is out fighting for his life" illustrates the amount of personal regard that he felt for his sales agents. It was easy for him to relate to them. He and other sales managers began their careers as sales agents and fostered strong connections with the men under them. His complaint to Benton shows his personal loyalty to the men in the field and all but dismisses their interference in the Arkansas election.

This letter also illustrates the depth to which the concern for adoptions at all costs grounded the evolving norms of sales agency: "It is difficult how some of our authors do not distinguish the relationship between their organization and our organization."[53] Personal loyalty united sales agents working within a company, but they felt no loyalty from the authors whose books they sold. Textbook authors and salesmen were occasionally antagonistic and seemed to operate at cross-purposes. Despite having different outlooks on the purpose of education, they depended on one another to bring their textbooks to the public.

According to Howe, the salesmen were much more aware of this dependence than were the authors. Actually, most authors did seem to recognize this, but they regarded sales agents as impediments rather than as colleagues. Most noticed that their publisher's sales force provided them with critical feedback, but they regarded the input as something to work around rather than to embrace. For trained judgment and a critical eye, textbook authors had one another to consult. Commentary collected in the field and passed through the editors to them usually required appeasement or even rebuttal.

No matter how strongly he felt, Howe did not write directly to the author. For all these publishers, the editorial department served as an intermediary. This had the effect of maintaining a distance between the two communities that allowed them to coexist. It also created a social bottleneck refracting information through the individual personalities who transmitted it from one side to the other. George W. Benton often played the role of diplomat, negotiating and balancing the needs of his sales force, on one hand, and his authors, on the other. Assistant Editor in Chief W. W. Livengood was perhaps more sympathetic to his company's sales force and regarded some of his authors, such as George W. Hunter, as more of a nuisance. Sales managers and agents he found more agreeable. In 1924, Louis B. Lee, the sales manager in Chicago, wrote to Livengood about a recent article in *School Science and Mathematics* that rated various biology textbooks: "I especially noted the rating of Ginn's Biology [Gruenberg's *Elementary Biology*], that is, the one that was rated very high. It seems to me the rater must have had a squint eye toward some publishers against some others." He alluded to the long history of competition between Ginn and the ABC and the practice of soliciting favorable reviews. Livengood, quite aware of this, wrote emphatically in the margin of this letter: "Quite right."[54] In other letters written at the same time, Livengood expressed doubts that Hunter would finish his next textbook anytime soon.

Editors served as the primary connection between sales agents and authors. In some cases, they served as the final arbiters regarding textbook content. More often, however, they were mediators who resolved disagreements between the two communities and tried to reach consensus. This placed them in an awkward position. Though they could make decisions regarding content and would bear ultimate responsibility for the quality of textbooks produced, they had to operate as much as possible in ways that would placate both their sales force and their authors. Often, this left little room for discretion and made them targets of blame that each side intended for the other. When textbook salesman and textbook author fought, the editorial desk was their de facto battleground.

Because of this, the personalities and temperaments of individual editors shaped the social flow of information, more so than the personalities on either side of the exchange. Editors of rival companies did not directly correspond, nor did they have large public networks to maintain with an outside world. Their professional contacts included their fellow editors within the company, the authors they worked with, the company's sales managers, its printing department, and its executives. Though exchanges were often cordial, much of the time the communications editors received were to address problems, either with a book's content or with some other person. Authors would occasionally correspond with sales managers, though this was not usual. George W. Hunter would occasionally meet with or write to sales managers of the ABC, usually when he was upset with an editorial suggestion and wanted to enlist someone's support against Benton, Livengood, or other ABC editors. This almost never succeeded.[55]

Out of such a complex of communities, high school biology textbooks came into being. Calling the process by which it happened *collaborative* is problematic. These communities coincided, but there was no sense in which they really worked together. This is in some ways true of most printed books. As Adrian Johns has written: "A large number of people, machines and materials must converge and act together for [a printed book] to come into existence at all. How exactly they will do so will inevitably affect its finished character in a number of ways. In that sense a book is the material embodiment of, if not a consensus, then at least a collective consent."[56] The early history of American biology textbooks seems to be especially complex. Given the system by which biology textbooks came into being, they are even less collective than other books. About the broadest agreement that these communities shared would have been a statement like, People should own biology textbooks. How they obtained them, what they said, why they should be read, and even what constituted a *biology* textbook were all questions on which there was no consensus. There may not even have been agreement as to why any of these questions were important. The textbook was less an embodiment of collective consent and more a chimera, the product of two divergent communities.

This chimerical entanglement of different genomes struggling to coexist in a single body aptly represents, not just textbooks, but biology itself. As the curricula of botany and zoology were commingled by the community of biology authors, the resultant discipline was endowed with social applications that reflected the authors' political ideology. Salesmen embraced the new texts as a way of creating a new market in biology, not to replace, but to augment botany and zoology. In some places, biology succeeded because of the content its authors had included; in others, it was adopted because of the salesmanship of agents. These overlapping interests sometimes obscured the reasons why the new biology was embraced—and why in some cases it was not.

If the different groups involved in creating the new biology had been more integrated, if textbook production had not been so complex, perhaps George Hunter would have received notice of the news from Mississippi that Charles St. John sent to Howe a few months before he started to revise the *Civic Biology* for a new edition: "I was very much interested in the Friday afternoon discussion on Science. Many of the Superintendents criticised General Science and strongly objected to evolution taught in Biology. Strong criticism was made of Ginn & Co. Biology now in use in Agricultural High Schools. This pleased me much."[57] St. John's letter may have been the first direct report the ABC had concerning the extent of the school antievolution movement, but it went unheeded.

It was around the same time that Ginn editor in chief Charles Thurber reached out to William Jennings Bryan to ask for advice on how to present evolution in the company's biology textbooks. The ABC's editors took no similar step, and, even though Hunter's *Civic Biology* had been controversial from almost the day it was published, they seemed to disregard any warnings about antievolutionism until John Scopes was indicted.

CHAPTER FOUR

Civic Biology and the Origin of the Antievolution Movement

On March 15, 1915, Walter G. Whitman, the science editor of the American Book Company (ABC), forwarded *Civic Biology* author George W. Hunter a letter from one of the company's salesmen. "One of the Boston High School Masters" had objected to the sexual content of Hunter's book, which discussed reproduction and the transmission of syphilis. In addition: "In Boston, the School Committee have issued an order that Eugenics is *not* to be taught." Eugenics—the science of human "improvement" through control over human reproduction—was a controversial subject, but most of the Boston complaints centered less on heredity (or even the notorious practice of eugenics through the sterilization of those deemed unfit) and more on sexuality. Despite his enthusiasm for the *Civic Biology*, this particular subject made even the textbook salesman uncomfortable: "Dr. Hunter has written a great book. But I am inflicting this letter in long hand to you because I don't feel that I can dictate what I want to say, and have said, to my stenographer who is a young lady of twenty-six years. Am I old fashioned in this? To how many of the girls in your office would you dictate the necessary letter to Dr. Hunter in the discussion of this matter?"[1]

Whitman avoided any indiscretion with his office staff by simply enclosing the original letter and dictating only an accompanying note to Hunter: "We have received some rather sharp criticism of your *Civic Biology* from Boston. This is undoubtedly of Catholic origin."[2] The next day, Hunter fired off a dismissal of the issue. The complaint was "hardly worth answering": "The writer has a view wholly biased and is evidently a fanatic on the matter of sex."[3]

The ABC's Boston office understood that these objections to the *Civic Biology* were religiously motivated. Nonetheless, it wanted Hunter to change the text. In July, Hunter was sent a list of suggested changes, including replacing discussion of syphilis as an example of contagious disease with a discussion of diarrhea and rephrasing an anecdote in the section on eugenics: no one objected to the description of "normal" or "feeble-minded" people (or to the diminutive contrast between a "man" and a "girl"), but they rejected the explicitness of the word *seduced* in describing how offspring came about.[4] The managing partner of the Boston office, Jarvis R. Fairchild, was enlisted to speak to Hunter. Editor in Chief George W. Benton informed Fairchild: "We have decided to hold up all work on his Laboratory Manual pending a settlement of these questions. This may be of some value in securing his reconsideration."[5] In September, Assistant Editor in Chief W. W. Livengood reported to Benton: "Hunter has come across and surrendered all the mooted points in his *Civic Biology*. Mr. Fairchild won him over . . . at a luncheon."[6] The fourth printing of the textbook included the relevant changes.

Perhaps it was the threat of holding up publication of the lab manual that won Hunter over, but Fairchild also tried to convince him that the ABC supported his book. As he told Benton: "I am anxious that everything in reason be done to meet [the Boston officials'] views, that no willful or even thoughtless offence be given them. But this is far from recommending that we publish no book that does not in every particular conform to Roman doctrine or practice." The stubborn, offensive textbook does not get adopted and earns no profit. But Fairchild did not think that avoiding controversy meant complete capitulation to political or religious pressure. As he related to Benton: "If you are going to omit these topics because our Catholic friends object, you will have to omit many others, or recast their statement. Will our Catholic friends accept any book based upon the doctrine of evolution—'the theory upon which (page 405) we today base the progress of the world?' "[7]

Ten years later, it was not Boston Catholics but Tennessee Protestants who objected to Hunter's textbook. And it was "the theory upon which . . . we today base the progress of the world," not its coverage of sexuality or eugenics, that made the *Civic Biology* the most infamous biology textbook in history.

THE CONTROVERSIAL *CIVIC BIOLOGY*

When the ABC first published Hunter's *Civic Biology* in 1914, its editors had no way of anticipating the Scopes trial. Even though evolution had its opponents, there was no effort to prohibit its being taught *in schools*. When the textbook first came out, the biggest criticism it faced came from those questioning whether its subject matter deserved to be called *biology*. The Cornell botanist George F. Atkinson complained: "It is not Biology. . . . To say Civic Biology is to my mind an anomaly and the word Biology seems to be put in to bolster up or to give tone to the introduction of certain industrial and domestic work."[8] For some scientists like Atkinson, *biology* implied a pure, abstract study of life. But, just a few years later, the University of Chicago biology professor Elliot Downing decried those textbooks that showed "little tendency to the industrial action or socialization of biology."[9] By the end of the 1910s, the change in high school biology to the civic model had become well-accepted among experts and was making its way into more schools as new textbook adoption cycles began. The *Civic Biology* was the first textbook to fully integrate the largely separate school subjects of botany and zoology. That overall approach—more than any single unit of the curriculum—gave rise to opposition to biology education, and that opposition found expression in the antievolution laws of the 1920s.

Stephen Jay Gould, Edward J. Larson, and others have argued that eugenics, as exemplified in Hunter's *Civic Biology*, was one of the primary reasons William Jennings Bryan came to oppose the teaching of evolution.[10] But there is no evidence in Bryan's writings against evolution (or in his available correspondence and papers) to suggest that he was particularly concerned with eugenics. It is entirely possible that some specific subjects in the *Civic Biology*—such as eugenics and sexual health—were controversial to some of the textbook's readers, but they were not related to the teaching of evolution, and they were not the biggest source of concern about the textbook. Most of the early objections that the ABC's editors received reinforced their perception that the primary complaint was, not individual topics, but the overall integration and organization of the *Civic Biology*.

The *Civic Biology* was the first of a generation of textbooks heralding a new way of talking about the life sciences in American high schools, and it ushered in the teaching of new courses on biology around the country. This was part of a larger wholesale change in how public education operated and who it served that spread across the United States in the early twentieth century. Civic biology symbolized the whole ideology behind education reform. More public schools were being built, teachers were becoming better trained and unionized, and public education was used to promote social change and create responsible citizenship. Objections to biology formed part of a larger outcry against these larger reforms.

At the time of its publication, the *Civic Biology* faced no direct competition from rival publishers. This was an unusual circumstance for the ABC's sales force. In many areas, the salesmen's biggest challenge was, not in arguing the merits of the *Civic Biology* as the best biology textbook, but in convincing educators and school boards that biology should replace or supplement the entrenched curriculum of botany and zoology. It may be because of these unique circumstances that the ABC's editors were willing to go to greater lengths to alter the book to accommodate sales—even when the author was reluctant.

The situation in Boston shows that, even when certain subjects like sexuality and eugenics were found objectionable in the 1910s, evolution was not. Of course, there were some religious groups that had already rejected evolution as a scientific theory. Several articles published in the widely distributed set of essays *The Fundamentals* (which spurred the Fundamentalism movement of the 1910s and 1920s) attacked "Darwinism" as antireligious.[11] But they did not advocate the legal recourse of banning evolution in schools. Instead, critics of education reform (some of whom may have had religious reasons for their opposition to compulsory schooling) co-opted the rhetoric of religious antievolutionism and turned it into a school movement. Complicated interactions between different religious ideas and Darwinian evolution were not new in the 1910s and 1920s. What had changed was the teaching of biology and the reform and expansion of compulsory public schooling. What was new was the publication of textbooks like the *Civic Biology* and the rise of state regulation of schools and textbook adoptions, which put these books into the hands of many students for the first time. It was in response to these contexts that the *school* antievolution movement emerged.

TEXTBOOKS AND THE DEVELOPMENT OF HIGH SCHOOL BIOLOGY

By the time of the Scopes trial in 1925, biological evolution had been well-known and accepted by virtually all American scientists for over half a century.[12] Yet the school antievolution movement was virtually nonexistent before 1920. There was no sudden realization that Darwin did not accord with the Bible. There were no new discoveries in either science or religion in the 1920s that led to antievolution legislation. What was new was schools teaching evolution, doing it in a politically charged way, and presenting it to a new population of students.

This was facilitated in the late 1910s and early 1920s by a trio of developments. The first was the creation of a new generation of biology textbooks that were organized around core principles of the life sciences and that intertwined applications of biology promoting certain cultural and economic worldviews. The second was a shift, particularly in the southern and western United States, away from local textbook adoption in favor of state-level regulation. This shift was largely motivated by factors unrelated to textbook content. The third was the expansion of compulsory high schooling into the rural South—particularly in Tennessee—which brought civic biology textbooks to students for whom other approaches to the life sciences were intended. The passage of the antievolution legislation challenged by John Scopes was a direct result of all these elements coming together. For all the talk about Darwinism and the Bible, Tennessee's Butler Act (named for the state legislator who introduced it) outlawing the teaching of evolution would not have been passed if there were not approximately four thousand ninth- and tenth-grade students in the state taking high school biology. In addition to the *Civic Biology*, the 1916 *Practical Biology* by Smallwood, Reveley, and Bailey had also been adopted by Tennessee in 1919.[13] It was slightly newer and less expensive than Hunter's book, but it was almost entirely unused in Tennessee. In 1924–25, around 90 percent of biology students used Hunter's 1914 *Civic Biology*, including the students in Rhea County Central High School, where John Scopes taught.[14]

The passage of the antievolution law in Tennessee had more to do with objections to compulsory education, state-level control, and the social values behind public education reform. The new field of biology exemplified these values because its textbooks were explicitly focused on the goal of preparing students for life as urban citizens in an industrial environment. The institutional structures that controlled public high schooling, that is, the way in which decisions were passed down from state to local levels, also contributed to the support for antievolutionism. The seizure of control of the schools by the state was seen as part of a larger effort to regulate local culture in ways consistent with the ideology embedded in biology education. To understand the antievolution sentiment that led to the Scopes trial, one must examine why the *Civic Biology* was written, why it was being used in a school in rural Dayton, Tennessee, in 1925, and why so many in the state objected to it.

The word *biology* dates back to the early 1800s, and its use in the title of textbooks came later in the century.[15] But these first "biology" textbooks largely treated plants and animals separately. In many ways, they were little more than separate books of botany and zoology stitched together in one binding.

There was some effort to supplement this format with some general principles. William T. Sedgwick and Edmund B. Wilson's *General Biology* exemplified this. Published for college use in 1886, it had a structure that made it well suited to the emerging field of biology as a rigorous science. As the authors announced in their preface: "Believing that biology should follow the example of physics and chemistry in discussing at the outset the fundamental properties of matter and energy, we have devoted the first four chapters to an elementary account of living matter and vital energy. In the six chapters that follow, these facts are applied by a fairly exhaustive study of a representative plant and animal."[16]

Sedgwick and Wilson's book contained two chapters dealing with "the biology of a plant" and four chapters on "the biology of an animal." A final chapter discussed the "outlines of classification" of both plants and animals. Though the discussion of energy, protoplasm, and cellular structure in the first chapters introduces some understanding of the unity of life, this theme was not carried throughout the book, and, in the books written for high school audiences that followed, the dichotomy between botany and zoology remained pronounced. As Philip J. Pauly described it, the biology course of the late nineteenth century was largely an "administrative solution" that "had little impact on the content of teaching."[17] The changes in pedagogy that made biology a subject in its own right did not come for another generation, with the textbooks of Hunter and his colleagues.

BIOLOGY—A COURSE PREPARED IN HELL'S KITCHEN

New York City was the epicenter of the new school subject of biology. The biology textbook authors Henry R. Linville, Benjamin C. Gruenberg, and George W. Hunter were all members of DeWitt Clinton High School's Biology Department, a place where, as Pauly describes, teachers "designed biology for upwardly mobile, immigrant Manhattan boys between the ages of thirteen and sixteen." That is, biology was first developed for students who were living in urban conditions—some in tenements, many first-generation Americans—and for whom encounters with zoology were largely limited to rats and roaches. DeWitt Clinton was "blocks from any play area and was bordered by piers, railroad yards and the Hell's Kitchen slum district."[18] The new biology was suited to this demographic, preparing students for life in industrial urban America, and focusing on public health and hygiene issues of particular importance in New York City. Pauly argues that it was the circumstances of their students that shaped these textbook authors' attitudes about education and about biology. The situation of DeWitt Clinton clearly demonstrated the need for the new biology, but in the cases of Hunter and Gruenberg (two of the most influential textbook authors of the era) views of the purpose and methods of biology education were largely formed prior to their arrival at Clinton.

George W. Hunter majored in biology at Williams College in Massachusetts, where he used Sedgwick and Wilson's textbook.[19] Like most liberal arts colleges, Williams offered separate courses in botany and zoology but was not large enough to support them in separate departments. A biology major at Williams would have to take courses in botany and zoology as well as some courses and workshops designed to appeal to both fields. After graduating in 1895, Hunter spent the following summer at Wood's Holl (now Woods Hole) Biological Station. Wood's Holl had been founded seven years earlier by Charles O. Whitman, chair of the University of Chicago's Zoology Department. Hunter matriculated to that department in October 1896.

Perhaps it was his experience in a department that focused exclusively on zoology that turned Hunter away from a career as a scientist, but his view of biology as a single, coherent field was in place long before he arrived at Chicago. His transcript from the university gives evidence of this. Other students in the Zoology Department at the same time as Hunter took almost all their courses in the department (with some taking a few courses in anatomy, histology, paleontology, or related fields in their later years). By contrast, Hunter enrolled in classes in botany, paleontology, and human physiology as soon as he arrived.[20] By the end of his second year, with several classes incomplete, he left without taking a degree. After a year teaching in Chicago, he joined the department at Clinton. Hunter's integrated view of the life sciences was quickly apparent as he immediately founded the extracurricular Biological Field Club (see fig. 2), which collected specimens of plants and animal life that were then used for teaching.[21]

Benjamin C. Gruenberg took a very different route to textbook writing. After graduating from the University of Minnesota, he moved to New York, soon enrolling in graduate school at Teachers College at Columbia. Around the same time, he and his wife, Sidonie Matsner Gruenberg (the future director of the Child Study Association of America), joined the Ethical Culture Society, a humanist, nontheistic organization.[22] After finishing his masters, Gruenberg joined Clinton's biology department.

FIGURE 2 Yearbook picture of the 1907–8 Biological Field Club at DeWitt Clinton High School. Club founder and faculty adviser George William Hunter is seated in the center of the second row. Courtesy DeWitt Clinton High School, Bronx, NY.

These relationships put Gruenberg into contact with many other individuals involved with education and social reform. Ethical Culture founder Felix Adler was a pioneer in advocating for schools to cultivate moral instruction independent of religion. In his 1892 *The Moral Instruction of Children,* Adler argued against the prevailing view "that morality depends on religious sanctions, and that right conduct can not be taught—especially not to children—except it be under the authority of some sort of religious belief." Anticipating the fears the school antievolutionists would express thirty years later, he noted that "the attempt to mark off a neutral moral zone, outside of the domains of the churches, is apt to be regarded as masking a covert design on religion itself."[23] Adler's work was influential in propagating the notion that, as nonreligious establishments, public schools could provide a kind of moral instruction.

Clarence Darwin Kingsley was a student at Teachers College at the same time as Benjamin Gruenberg and soon after graduation also joined the Ethical Culture Society.[24] Gruenberg and Kingsley were both active in committees looking at the state of education in New York City. In 1911, Kingsley served on a committee chaired by Gruenberg that evaluated the role of schools in promoting child welfare.[25] In 1913, Kingsley was appointed chair of the National Education Association's Committee for the Reorganization of Secondary Education, whose reports were critical in shaping the role of high schools in American life in this period. A committee on science education issued its report in 1920. The subcommittee was chaired by Otis W. Caldwell, who cowrote a general science textbook published by Ginn in 1914 and would join the Teachers College faculty in 1917.[26] On the committee were also the high school biology teacher and textbook author James E. Peabody and former ABC science editor Walter G. Whitman.[27]

Henry R. Linville was another important person who connected the development of science education with a wider political progressivism highlighting the role of schools in social reform. As the first chair of Clinton's Biology Department, he was responsible for the development of the school's course of study. He recruited Benjamin Gruenberg to the school and continued to support Gruenberg's career even after both had taken jobs elsewhere.[28] Linville was, like Felix Adler, involved in the founding of the ACLU, and he was responsible for its involvement in combating Tennessee's antievolution law.[29] One of the first causes the nascent ACLU took up was advocacy for conscientious objectors to the Great War, a group that included Linville. After the war, New York teachers were compelled to sign oaths of loyalty, which Gruenberg did under protest.[30] The state of New York eventually investigated claims that the principal of DeWitt Clinton High School had attempted to purge the school of those who refused to sign loyalty oaths, especially those assumed to have socialist sympathies. Linville testified on behalf of those dismissed from the school.[31]

The issue of protecting teachers from discriminatory treatment by the state or by their principals also led to the formation and growth of teachers' unions in New York. Linville helped establish the New York City Teachers Union and went on to serve both as its president and as the editor of the *American Teacher*, the newsletter for the national American Federation of Teachers.[32] The AFT was also another mechanism by which the education theories of John Dewey were spread to a growing population of professional, well-trained teachers. In 1904, Dewey joined the faculty of Teachers College and the Columbia University Philosophy Department. In addition to teaching and working with many of the city's science education leaders, Dewey served as vice president of the New York City Teachers Union under Linville's presidency.

Gruenberg was connected to each of these individuals and was involved in nearly all their activities. He gave lectures at the Ethical Culture Society, was an officer in the New York City Teachers Union, served on national committees with reformers like Kingsley, and frequently corresponded with fellow science textbook authors like Linville, Peabody, and Caldwell.

In the minds of these individuals, there was a connection between the values of the movements they supported and how pedagogy itself worked. It was not the case that reformers of science education just happened to be interested in progressive solutions to economic, public health, and social problems; the felt need for civic reform and science education came from a common ideology. In the reformers' view, schools were capable of doing more than just training future laborers. The Committee for the Reorganization of Science Education, chaired by Otis Caldwell, explained: "The members of a democratic society need a far greater appreciation of the part which scientifically trained men and women should perform in advancing the welfare of society. Science teaching should therefore be especially valuable in the field of citizenship because of the increased respect which the citizen should obtain for the expert, and should increase his ability to select experts wisely for positions requiring expert knowledge." This included the teaching of socially responsible ethics that *could* be presented in public schools because they were not religious in orientation: "Science study should assist in the development of ethical character by establishing a more adequate conception of truth and a confidence in the laws of cause and effect."[33] This was a specific proposal that echoed Felix Adler's exhortation that teaching ethics could be separated from teaching religion.

The basis of those ethics and of improved urban conditions was to be found in social organization and scientific advance. Courses teaching this reorganized science were designed to speak to the real-life experience of students. Science education was necessary to social progress because science could deal with examples of social needs that students encountered every day. It is possible that some individuals involved in this movement developed this method in contrast to religion or to religious styles of learning. John M. Heffron suggests that Caldwell's emphasis on science education's need to speak to everyday life was born of an adverse reaction to his rote memorization of the catechism as a youth.[34] Caldwell's views on this were expressed in the committee report of 1920: "The new science should also develop direct, effective, and satisfying methods of solving problems. If these methods are to be of wide use outside the school, they must be formed through and firmly associated with the kinds of experiences that arise in common needs. Real situations and good methods consciously and constantly applied with satisfying results are necessary for this purpose."[35]

These recommendations were instrumental in the creation of a general science course. As John L. Rudolph notes, Caldwell's general science book (written with William Eikenberry) was advertised as teaching students *how* to think.[36] This was seen as one of the main virtues of the general science class, providing the basis for further study in biology, chemistry, and physics.

Evolutionary theories were central to this. In particular, the applied evolutionary notion that the human race could advance through social efforts was highlighted in claims about science's contributions to progress. This thought underlay eugenic practices, but it also gave rise to developing sanitation and improving the quality and safety of food and water. All these contributions were included in Hunter's *Civic Biology* as examples of making the environment better for mankind.[37]

While the recommendations to reform science education applied to each of the school disciplines, it was in biology that they were most integrated. The living world was not the only part of nature that led to "the kinds of experiences that arise in common needs"; chemistry and physics could lay claim to that as well. But more attention was paid to biology in part because education reformers looked to biology to explain what student minds were like and what they were capable of learning. Evolution was not just something students were exposed to as a subject in the life sciences; it was a way for teachers to understand student minds and to teach them how to think. Biology was not just the *subject* taught; it was the basis for understanding how teaching and learning *worked* and at what stage of student development new subjects could be introduced.

One version of this evolutionary understanding of mind was the idea of *recapitulation*. Though there were several associated theories that went by that name, the one most influential among educators was articulated first by Herbert Spencer. In his 1860 *Education: Intellectual, Moral, Physical*, Spencer noted: "Education should be a repetition of civilization in little. It is alike provable that historical sequence was, in its main outlines, a necessary one; and the causes which determined it apply to the child as to the race."[38]

In this view, the child mind is not sufficiently developed to comprehend the abstract and complex intellectual achievements of civilization. Children begin with a very basic moral and intellectual sense and develop new cognitive abilities as they age. Education, therefore, must be sensitive to that development. Felix Adler, in his *Moral Instruction of Children*, "assumed as a starting-point the idea that the child rapidly passes through the same stages of evolution through which the human race has passed." He outlined the different kinds of moral education of which the child is capable at successive stages: "In the primary course, the object has been to train the moral perceptions; in the grammar course, to work out moral concepts and to formulate rules of conduct. . . . In the advanced course . . . we shall have to reconsider from a higher point of view many of the subjects already treated, and in addition take up such topics as the ethics of the professions, conjugal ethics, etc."[39]

John Dewey cautioned against putting too much credit in the recapitulationist view of education in his 1916 *Democracy and Education*: "Embryonic growth of the human infant preserves, without doubt, some of the traits of lower forms of life. But in no respect is it a strict traversing of past stages." Progress requires more than mere recapitulation; in Dewey's view, individual development must take shortcuts to reach the developmental stage of the present age: "The aim of such education is to facilitate such short-circuited growth. . . . To ignore the directive influence of the present environment upon the young is simply to abdicate the educational function."[40] The goal of education is to promote social well-being by pushing individuals past the apparent extremes reached by their forebears. As such, education promotes social progress, not only individual enrichment. Despite Dewey's rejection of recapitulationism, his view was also evolutionary. Some have described it as "neo-Lamarckian."[41] Too slavish a devotion to recapitulation might lead one to the conclusion that social progress could not overcome innate biological limitations.

Historians of education have framed the intellectual development of schooling in the early twentieth century as a split between the theories of Dewey and those of his Teachers College colleague Edward L. Thorndike. Both camps made use of evolutionary accounts of mental development, but part of what divided them was different beliefs in the aims of schooling and the kind of social progress the schools could contribute to. Though both sides of this debate are complex and group together the views of many individual scholars, the distinction is often represented, on the one hand, as a focus on standardized instruction intended to prepare the masses for contribution to society (Thorndike) and, on the other, a focus on cultivating independence of thought so that individuals contribute more robustly to a democratic society (Dewey).[42] Despite some differences, the psychologist and educator G. Stanley Hall largely agreed with Dewey's view of the aims of science education. (Thorndike described Hall as a man "whose doctrines I often attack, but whose genius I always admire.")[43] As the president of Clark University, Hall had hired Charles O. Whitman to lead the school's Biology Department in 1889. Hall was perhaps the most important proponent of psychological recapitulation as an account of the learning process. He decried the "order of teaching science as commonly practised, especially in the secondary school," which "does violence to the nature of the child" through its focus on memorization and classification. Hall was one of the first to suggest that "in biology the themes of heredity, variation, recapitulation, natural and artificial selection, the struggle for existence, development histories, lessons from palaeontology—all such large themes—form the most practical science for the secondary school."[44] The emphasis on these biological themes in science education was seen as an appropriate response to theories of the developing mind that accentuated the evolution of a democratic society.

The idea of social evolution, particularly Spencer's, appealed to Edwin Ginn.[45] Perhaps that explains why his company published the first general science textbook, Caldwell and Eikenberry's. The book came out shortly after Ginn's death in 1914 but had begun under his guidance. At the same time, Ginn's adversary, the ABC, was guiding Hunter's *Civic Biology* to market.

George W. Hunter was more disconnected from the larger political debates about the role of education and scientific progress than his colleagues Gruenberg and Linville. But his studies had given him a more synthesized view of botany and zoology as an integrated subject, a view that he brought with him to New York. Nonetheless, the first two textbooks Hunter wrote, the 1907 *Elements of Biology* and the 1911 *Essentials of Biology*, were still basically arranged along the old botany-zoology-physiology structure of biology textbooks published in the previous century. The last chapter of the *Essentials*, however, was called "Health and Disease—a Chapter on Civic Biology." It discussed issues of sanitation, food safety, and infection but made little effort to connect these issues to discussion in previous chapters.[46] Hunter was already teaching at DeWitt Clinton High School along with Gruenberg and Linville when these two textbooks came out, so, if there was a change in biology teaching at Clinton, it was not yet apparent in his textbooks.

It is likely that the science editor Walter G. Whitman (who would later serve on the committee Caldwell chaired to reorganize science education) encouraged the ABC to invest in textbooks to fit the new curricular recommendations that were being developed. Whitman later coauthored a series of books on general science with George Hunter, the first being *Civic Science in the Home* in 1921, followed by *Civic Science in the Community* in 1922.

These two works (republished in 1923 under the single title *Civic Science in Home and Community*) had the word *civic* in their titles to emphasize their continuity with Hunter's *Civic Biology*.

CIVIC BIOLOGY AS BIOLOGY FOR CITIES

Hunter's *Civic Biology*, with its focus on such issues as quarantine, alcohol, food safety, and the improvement of human society (including a substantial section on eugenics), was geared toward America's growing cities. This was made clear in the book's preface: "This book shows boys and girls living in an urban community how they may best live within their own environment and how they may coöperate with the civic authorities for the betterment of their environment."[47] This new curriculum began in New York City, giving what Philip Pauly called the "metropolitan origins" of civic biology.[48]

The 1914 *Civic Biology* combined Hunter's sense of integrating plants, animals, and humans into a coherent science of life with the ideology of scientific progressivism and the role of education in social progress. The ideology behind civic biology as a movement was one that saw schools as the entry point for the improvement of all human society. As Felix Adler suggested, schools could be used to teach a morality that was not grounded in theology. As Dewey had explained, that morality was social, aimed at the cultivation of citizenship. Kingsley's, Whitman's, and Caldwell's efforts to reorganize secondary education were predicated on these ideals as the purposes of education. Part of creating an environment in which such social learning could occur involved the use of lessons that touched students' everyday lives and that had teachers who had the ability and the autonomy to go beyond rote instruction.[49]

With the *Civic Biology*, the old framework of biology was overturned to emphasize the human use of knowledge as the central aim of the study of life. The botany/zoology bifurcation was done away. Diverse issues such as the environment, economic use, and heredity and variation were discussed for plants and animals simultaneously. Hunter described this reorganization as the way to engage the real-life experience of the urban student:

> Children in rural schools wish to study different topics from those in congested districts in large communities. The time has come when we must recognize these interests and adapt the content of our courses in biology to interpret the *immediate* world of the pupil.

> With this end in view the following pages have been written. This book shows boys and girls living in an urban community how they may best live within their own environment and how they may coöperate with the civic authorities for the betterment of their environment.[50]

Hunter and his colleagues teaching biology in New York City restructured their courses around the social and economic uses of biology, abandoning disciplinary distinctions between plant, animal, and human life in favor of a course that was organized around central concepts common to living things.[51] These included heredity, cellular structure, metabolism, evolution, public health, and the use of biology to improve social conditions.

The *Civic Biology* had the virtue of novelty and the weight of the ABC's sales machinery behind it. The power of the book trust at its height assured the book's success. In ways the publisher could not have foreseen, *Civic Biology* may also have been helped by some teachers embracing the civic approach to science because of the reputation and credibility of people like Henry Linville, who had a national voice as the editor of the *American Teacher*. The role of science education reformers in the early unionization of teachers gave their views on pedagogy greater distribution to teachers across the country and contributed to demand for the civic approach to science. (Although, with textbooks being chosen by school boards, which frequently opposed teachers' unions, this may also have hurt the cause of civic science in some places and likely was a reason why Linville's own textbook was not as widely adopted.) With the first (and for a while only) biology textbook of its kind, the ABC reaped the benefits. Despite early complaints, by the end of the decade the *Civic Biology* was the best-selling textbook in biology. It was even adopted in Boston, the home of the rival publisher Ginn and one of the first places to voice an objection to Hunter's textbook. Other publishers quickly recruited biology teachers to write textbooks for them.

With the advent of civic biology as the new way of teaching the life sciences came a new generation of textbooks. Instead of core themes of biology being treated in a separate first section, they became the organizing principles of the *Civic Biology* and subsequent textbooks. These principles were integrated into the discussion of specific classes of plants and animals and were presented with uses that illustrated their centrality and their real-life importance. Civic biology intertwined the new understanding of the life sciences that had compelled the shift away from the separate study of botany and zoology with the social applications of that knowledge. One could not teach biology from these textbooks without teaching its applications. Indeed, in the minds of these textbook authors, the *point* of teaching these concepts was their application to the human world.

Thus, with cellular theory came discussion of microbial understandings of disease and the need for public sanitation, quarantine, and hygiene. With metabolism came discussion of the effects of alcohol consumption and the importance of a healthy diet. With heredity came discussion of reproduction and sexual health. And with evolution came eugenics and the cultivation and improvement of plant and animal species for human use.

Civic biology was never just about teaching evolution *as a subject*. The principles behind civic biology as a kind of teaching were evolutionary in nature—the evolution of a society aided by scientific progress, the evolutionary development of young minds at an age when they could be taught abstract reasoning and eschew rote learning. Civic biology was developed, not only to teach students the theory of evolution, but also to bring human evolution about. Hunter's discussion of the common descent of animal species or the historical progression from primitive to modern humans was the least important way in which the book presented evolution.

The efforts of textbook publishers, the influence of Dewey and other pedagogues, the spread of teachers' unions, and the rise of immigration, urbanization, and industrialization in the United States all helped integrated civic biology become the way to teach life sciences. In 1915, 6.9 percent of all science course enrollments in public high schools were in biology. This figure had grown to 8.8 percent by 1922 and 13.6 percent by 1928. The percentage of students taking stand-alone courses in botany and zoology declined substantially as those courses were replaced by biology. At the same time, the overall number of science course enrollments nearly doubled as more schools opened and existing schools offered more science courses.[52]

While the leading advocates of expanding compulsory public education wielded the science textbook as an instrument of social reform, the textbook industry itself was also gleefully supportive of a political movement that would mean more students and more sales. A nationwide consensus on school subjects (even if a national standard curriculum was impossible) also meant investing in the development of fewer titles without sacrificing sales. Publishers did realize, of course, that the same book could not be sold to every district. Civic biology had, not only "metropolitan origins,"[53] but also northern origins, industrial origins, organized labor origins, and progressive origins. Aware that some people would be skeptical of civic biology, some publishers did not abandon the older way of teaching the life sciences, simply repackaging it for a more specific audience.

TEXTBOOK PUBLISHING AND THE URBAN-RURAL, TWO-MARKET APPROACH

Publishers recognized that civic biology textbooks were written for urban audiences. Walter Whitman once inadvertently referred to Hunter's book as "A Civic Biology Presented in Problems for City Schools."[54] Sales catalogs also explicitly described the textbook as geared to the urban student. Not wishing to lose rural markets, some publishers updated their older biologies in the early 1920s or issued new ones intended for rural markets. In some cases, they marketed different biology textbooks by the same authors. Hunter's *New Essentials of Biology* came out in 1923 and was marketed alongside the *Civic Biology*. Smallwood, Reveley, and Bailey's 1920 *Biology for High Schools* was sold alongside their 1916 *Practical Biology*.

The *New Essentials* was a revision of Hunter's 1911 *Essentials of Biology*, which had been written under a more traditional scheme that split biology into separate units of botany and zoology. Though the *New Essentials* was not quite as bifurcated as the original, it retained much of the ordering of the *Essentials*, with the first chapters dealing with plants, one chapter relating plants and animals, and then chapters covering zoology. After these came chapters on human physiology and a single chapter on civic biology. The preface of the new book took note of the developments in pedagogy that had occurred with the *Civic Biology* and its imitators and attempted to position the *New Essentials* as having taken advantage of the new developments in science teaching. Despite the book's organization, Hunter's preface suggested: "The data should be treated from the *biological* standpoint, not that of botany, zoölogy, or human physiology. Ideally, we might take up general principles and draw from the great storehouses of plant, animal, and human biology to illustrate each principle before going on to the next."[55] In effect, while ABC salesmen wanted a textbook for an audience not willing to accept the pedagogical synthesis of civic biology, Hunter encouraged teachers to draw these connections for themselves.

As a result of Caldwell's committee report, in the 1920s urban schools began shifting biology to the second year of high school, making room for a new first-year course in general science.[56] By this time, Hunter had left New York and was a professor of biology education at Knox College in Illinois. While there, he published studies on the development of this new course and the shift in grade placement.[57] But general science did not take hold in most rural schools until much later.[58] Consequently, the idea that science education should be different for urban and rural audiences was further entrenched by the publication of the *New Essentials*: "The plan of this book recognizes first-year biology as a *science* founded upon certain basic and underlying principles. These principles underlie not only biology, but also organized society."[59] Incorporated into the urban-rural split manifested by the disparate textbooks used to teach biology, therefore, was also a division over the utility of general science. John L. Rudolph describes general science first taking root in "schools where countless children toiled to learn the methods and applications of science in their unsettled urban communities."[60] The general science movement not only had urban origins but also shared in the same social goals as civic biology. By shifting civic biology to the second year in urban schools while keeping rural biologies at a ninth-grade level, publishers reinforced the civic nature of general science and exacerbated the education differences between urban and rural America. When Hunter and the ABC began to discuss revising the *Civic Biology* in 1924, one goal was to make it into a tenth-grade book instead of a ninth-grade one.[61]

Ironically, textbooks like the *New Essentials* were most successful in the rural quarters of states with large cities in the Northeast and Midwest. This was partly because those states had more widespread rural education, but it was also because most of the South and West regulated textbook adoption at a statewide level. In many of these states, rural education was not yet widespread. In states where schools, experienced teachers, and superintendents from rural areas were few in number, their interests were often underrepresented on textbook commissions.

Statewide regulation of textbooks may also have been necessary in places where the very existence of compulsory schooling was contentious. As Charles A. Israel has shown, public education was very controversial in parts of Tennessee in the years leading up to the Scopes trial, and the debate over schools had religious implications. Some evangelical Protestants feared that public schools would contribute to the secularization of their communities, while other religious groups saw education as the fulfillment of a moral calling to improve society. These school advocates encouraged religious participation in public education. Their support was part of the political compromise that allowed earlier compulsory attendance laws to be passed in the state.[62]

Textbook regulation by states was also seen as a progressive response to the marketing practices of the *book trust*, as the ABC was frequently called by its critics. Although state-level regulation of textbooks was not originally based on a desire for state-controlled content, state-level uniform textbook adoption in the South forced people in charge of education to make decisions about what kind of biology textbooks to use. They could either choose a more traditional rural textbook—which would not teach the essential concepts of biology as well and were sometimes seen as inferior in their approaches to learning—or they could adopt a civic biology textbook that included a social agenda that some would find objectionable. Textbook commissions were compelled to confront their states' changing demographics and consider the role of education as an instrument of social change.

These were not issues unique to biology. The "rural school problem" had long attracted the concern of education scholars and social reformers who observed a growing disparity between urban and rural schools. Some questioned whether rural cultures were inherently antithetical to a good education, while others saw the differences as consequences of poor school funding and administration. Small school districts in rural settings had neither the resources nor the administrative mettle found in schools in the cities.[63] Shifting textbook control to the state was one response. As Tracy Steffes describes it, the rural school problem was presented as "a microcosm of larger national anxieties about the place of rural life in an increasingly urban-industrial nation." It raised questions about the long-term future of an American culture rooted in "the children of largely native, white yeoman farmers" challenged by the influx of immigrants to growing American cities.[64] There were some efforts to urbanize rural schools through consolidation of small districts, and solutions were also sought that recognized the unique education needs of the country. When public education in the South expanded to more rural areas, choosing a uniform textbook for a state as a whole proved more troublesome.

ANTIEVOLUTIONISM AND SCHOOL REFORM IN TENNESSEE

Austin Peay was a progressive Democrat and a Baptist, and, with his reelection in November 1924, he focused his political agenda on public education. In hindsight, the most famous law passed by the Tennessee legislature in 1925 is undoubtedly the Butler Act. At the time, however, the most important matter before the legislature was the general education bill. This bill was intended to expand education throughout the state, establish a minimum eight-month school year, and provide for the establishment of at least one high school in every county in the state. The bill would more than quadruple the amount of money spent by the state on education.[65]

This was an ambitious bill, intended to have dramatic effects. Tennessee ranked near the bottom among the forty-eight states and the District of

Columbia in terms of literacy. The education bill that Harned and Peay drafted met with resistance in the legislature, where concerns over the social implications of public education, as well as its expense, led to protracted debate and significant compromise.

One particular region of Tennessee whose citizens were largely opposed to the bill was the Cumberland valley, a rural, mountainous area of Middle Tennessee. In 1924, these citizens elected John Washington Butler to the State Senate. Jeanette Keith has observed that Butler's antievolution bill was intended as a protest against continued state intrusion into the cultural life of the region: "Farmers of J. W. Butler's generation believed self-sufficiency a worthy goal; as Butler said, they wanted the 'right kind' of education for their children, the sort that would make youth 'thankful that they live on a farm, where they can make what they eat and eat what they make.'" Butler had religious objections to the theory of evolution, but those objections found expression only in legislation directed against the schools, which could not have happened before the general education bill was proposed. The content of what children were taught in schools could not have been controversial until children were required to attend school. The general education bill meant the sudden prospect of compulsory high schools in regions of the state where few people had even had the option of a high school education. Owing to the perception (reinforced by biology textbooks) that high schools taught a way of life geared toward urban living, not rural culture, legislation concerning the schools was of great concern to the people of the Cumberland valley. Keith concludes: "If the Tennessee Monkey Law held any symbolic meaning on the local level, it was a regional reaction to decades of reform culminating in a loss of local control over education."[66] This sentiment was echoed by Governor Peay, who, when he signed the antievolution bill into law on March 23, 1925, called it "a distinct protest against an irreligious tendency to exalt so called science." Echoing arguments Bryan had made earlier, Peay asserted that "the people have the right and must have the right to regulate what is taught in their schools." At the same time, however, he reassured the law's opponents: "Probably, the law will never be applied."[67]

There had been other attempts—somewhat successful—to legislate the place of evolution in the schools. In 1923, Oklahoma briefly banned the use of textbooks containing evolution, and, in 1924, Florida passed a nonbinding resolution opposing the teaching of evolution. But the Tennessee law introduced by Butler did not regulate the content of textbooks. Instead, it prohibited the actions of teachers and, in doing so, made possible criminal prosecution for violation.

There have been several explanations put forward as to why the Tennessee legislature passed Butler's bill and why Governor Peay did not veto it. Kenneth K. Bailey described how the speaker of the state Senate intervened to enable the Butler Act's passage after a similar bill had already been rejected earlier in the session. Bailey observed that Peay received numerous requests from both supporters and opponents of the bill.[68] Edward J. Larson suggests that, in signing the bill, Peay was not only following his own inclination but also following the majority will of the people.[69] The legislature was also considering a bill to provide over a million dollars for new buildings and equipment at the University of Tennessee.[70] Larson suggests that some people quieted their opposition to the Butler Act in order to gain support for the university. But, for Peay, the most important matter was the general education bill. As Paul K. Conkin notes: "One clear reason that Peay signed the Butler Act was his desire to gain legislative support for his reform package."[71] The chronology of events surrounding Peay's non-veto makes that connection between the antievolution law and the general education bill all the more apparent. A first version of the education bill was introduced in the State House of Representatives on February 6, 1925. On February 9, it was referred to the Committee on Education, which never voted on it. The day after the legislature sent the antievolution bill to the governor, a new version of the education bill was introduced and sent to the same committee that had assured the death of its predecessor. Many supporters of science education, and even some religious leaders, encouraged the governor to veto the antievolution provision, but, with the fate of the general education bill in the hands of the legislature, he signed Butler's bill on March 23. Two days later, the general education bill was recalled from committee, and debate began. After more than a week of discussion, during which 156 amendments were considered, the bill passed on April 2.[72] The Senate passed the bill with an additional 121 amendments April 15, and the final version of the bill was sent to the governor on the last day of the legislative session.[73] The general education bill became law only through months of politicking, compromising, and negotiating by Peay and Harned. Peay's signing of the Butler Act was one part of that process.

Peay's non-veto of the Butler Act also had other political effects that may not have been intended, but it certainly benefited the governor in that it gave him the ability to set education policy for the state. The enactment of the antievolution law led one Republican member of the Tennessee Board of Education, Mrs. C. B. Allen from Memphis, to resign in protest.[74] Peay and Harned filled the seat with a Democratic woman.[75]

While it is a stretch to suggest that Governor Peay had Allen's resignation in mind when he signed the bill into law, it is curious to note that just three weeks earlier she had refused to resign her position when opponents claimed that she was spending the majority of her time living in Chicago.[76] Explaining her newfound reasons for her resignation, she wrote the governor: "While this law cannot permanently harm either religion or science, as the search after Truth will go steadily onward, despite this feeble and futile attempt to curb thought . . . I believe that by remaining a member of the State Board of Education of Tennessee . . . I should be giving my silent consent and approval of this Anti-Evolution Law."[77] Newspapers reported that Allen had been compelled to leave because the general education bill would have barred her from serving on the state school board as a nonresident.[78]

In exchange for expanding compulsory public education into rural Tennessee, Peay consented to what he thought would be a symbolic protest against state control of school content. The great number of amendments and the amount of time spent on debate over the general education bill show how politically fraught school reform was in Tennessee and how hard Peay and Harned had to work to bring it about. To some extent, their political maneuvering began in 1923 with the plan to postpone textbook adoptions until after the next election. Once the second version of the bill had been introduced in the 1925 legislature, the defining moment in its passage came when it was rescued from the Committee on Education, where the previous bill had died. This occurred only after the governor accepted and signed the antievolution bill. Without this, there may have been no education reform at all.

Many rural residents saw the expansion of public education in Tennessee as an attempt to change their culture and to instill foreign values. Civic biology taught students to prepare for a life away from their traditional upbringing. Consequently, parents took exception to the *presence* of biology as well as to its content. The fact that the books taught the historical development of species was a small concern. The overall discipline of civic biology and the presence of new schools intended to bring social progress were much more objectionable.

By 1922, when William Jennings Bryan's front-page editorial in the *New York Times* heralded antievolutionism as a *school* movement, civic biology textbooks were already firmly entrenched.[79] Until evolution appeared in the textbooks, there would have been no reason to ban the teaching of it. Focusing on evolution as a subject *within* biology and pointing to religious objections to Darwin in particular gave a voice to those fearing that science education was integrated with values antithetical to their culture. With the passage of Peay's 1925 general education bill, the loss of local control over school content was complete. Passage of that bill relied on the symbolic protest over school content represented by the antievolution law.[80]

A TEXTBOOK VIEW OF THE SCOPES TRIAL

The development and distribution of a new type of biology textbook was influential in the rise of antievolutionism as a school movement and the legislation it influenced. It is even possible that Tennessee's decision to outlaw *teaching* evolution rather than prohibiting the use of textbooks containing evolution was influenced by the textbook industry. But publishers' impact on antievolutionism in Tennessee did not end with the passage of the Butler Act. Textbooks and their publishers played an instrumental role in the Scopes trial itself and in the public perception of the trial's origins and aims afterward.

The night before the trial began, Benjamin Gruenberg received a telegram asking him to testify as an expert witness in Dayton: "We of the defense would be delighted to add you[r] authority to our position."[81] The telegram was signed with Clarence Darrow's name. Marcel Chotkowski LaFollette has suggested that this telegram was one of many sent on Darrow's behalf to scientists and potential experts by the Science Service news organization.[82] But Gruenberg appears to have been the only potential witness directly involved in high school biology education. Many of those contacted did not go to Dayton.

On Friday, July 10, 1925, the first day of the trial, Gruenberg wrote to one of his editors: "This is worth getting into, and it should not take more than a few days."[83] He also sent a telegram back to Darrow: "I should consider it a privilege to be of assistance in this very important trial."[84] At the Ginn offices, word traveled from Boston to Tennessee and back to New York, and by Saturday Gruenberg received an urgent telegram from L. D. Roberson, the partner in charge of Tennessee at Ginn: "I strongly urge you not to go to Dayton to testify. It will kill your book in Tennessee and throughout the South and greatly injure Ginn and Company. I trust you will neither go nor get your name mixed up with it."[85] On Monday, Gruenberg wrote back to Darrow, telling him: "I find that it will be impossible for me to get away at all for several weeks, as material from the press that I cannot shift to others is urgently awaiting my attention."[86]

Given Gruenberg's previous battles for the rights of teachers and for textbook authors, his attraction to the fight in Dayton seems wholly consistent.

Gruenberg was a prolific writer and editor; he may really have had urgent work.[87] But the rapidity of his change of heart suggests that Ginn's pleading had an effect. More than an indication of Gruenberg's attitude toward the Scopes trial, the intervention by Ginn speaks more directly to the publisher's pragmatic attitudes. It was not long before Ginn sales agents were using Hunter's *Civic Biology*'s association with the Tennessee case to discredit the ABC and promote Gruenberg's own textbook.

Gruenberg's new (March 1925) book, *Biology and Human Life*, had just been added to the list of adopted books in Tennessee. It joined the Macmillan-published *Biology and Human Welfare* by James E. Peabody and Arthur E. Hunt, which had been adopted by the Tennessee state textbook commission on June 16, 1925—after Scopes's first indictment but before the trial itself.[88] One member of the commission explained: "[We] didn't want to cause any scrap so we cut out Hunter's biology, which teaches evolution, and adopted one that makes no reference to it."[89] The defense attorney John R. Neal protested the adoption of a new biology textbook while the Scopes trial was still pending and wrote to Governor Peay. Because "said case will test the constitutionality of the Anti-Evolution Act passed by the recent Legislature," he demanded "postponing the selection of the Text Books on Science until after the final determination of the case." Neal, who had run against Peay in the Democratic gubernatorial primary in 1924, threatened consequences if the request went unheeded: "If the Commission should not postpone the selection . . . it would be, in our opinion, a violation of the law . . . and criminal prosecution will be immediately initiated."[90] Peay testily informed his commissioner of education, P. L. Harned: "The State text book Commission will adopt a book on science without any regard to Mr. Neal's Letter."[91] And Hunter's aging textbook was replaced.

The sales representatives at Macmillan had tried to represent the book as compliant with the newly enacted "Monkey Law," but the authors themselves were surprised to find it adopted. Arthur Hunt stated: "The book really does not treat of evolution. It is designed for students of from thirteen to fifteen and evolution is not an elementary subject. Of course, both myself and Mr. Peabody are evolutionists."[92] Peabody and Hunt may have been personally opposed to antievolutionism, but they were less willing to get involved in textbook fights than some of their colleagues. Their muted response to antievolutionism evokes comparison with their decision not to join in the letter of protest concerning Arthur Clement's textbook in 1924.

Gruenberg's invitation to testify and his brief period of acceptance say less about the trial itself than they do about the role textbook salesmen, publishers, and authors played in shaping how the trial was perceived.

Even if Gruenberg had gone to Dayton, his participation would not have affected the legal outcome of the trial. None of the defense's expert witnesses were permitted to testify about evolution. Nonetheless, many of the experts were interviewed by the media covering the trial, and Gruenberg's presence might have called attention to the trial's importance to education policy. Though Ginn's concern was centered on its own immediate sales in Tennessee, the company's veto of Gruenberg's participation also facilitated the representation of the trial as a debate over science and religion instead of as an education issue. In the aftermath of the trial, textbook publishers would continue to reinforce that perception in the ways in which their biology textbooks were presented as responses to the Scopes trial.

The publishers were far from alone in promoting science and religion as the central issue in Dayton. Without Gruenberg, or, indeed, any expert witnesses permitted to testify, Darrow famously called William Jennings Bryan to the stand. The complex issues that led to the Scopes trial were lost in the polarizing portrayal of science and religion in irredeemable conflict and a distorted debate over the "literal" truth of the Bible.

Historians exploring the complexity of science and religion have argued with compelling evidence that religion does not always oppose science.[93] Religious responses to evolution were diverse, ranging from rejoicing to rejection, and not every antievolution argument was religious. The Scopes trial's participants reconstructed the origins of the school antievolution movement to suit their own needs. Accepting this framework has resulted in obscuring some of the important trends in biology education and textbook marketing that contributed to the event. It has also enshrined a sense that the trial, or something like it, was inevitable from the moment someone first put the *Origin of Species* on a bookshelf next to the Bible. The story of the trial and its origins is much more complex, as is the story of religion and antievolutionism. Religion may have been used to *justify* school antievolutionism, but it was not the primary cause of it. Indeed, given what many rural southerners experienced of education reform and how they perceived civic biology and its textbooks, it is easy to understand their skepticism about the motivations of new schools and their ambivalence toward evolution. In 1915, Boston schoolmasters objected to Hunter's *Civic Biology* for reasons that were quickly characterized as religious. But the idea of religious antievolutionism was so remote that it was put forward as an absurdity. Ten years later, opposition to Hunter's book specifically concerned evolution. The resulting antievolution law, grounded in appeals to the Bible, led to the Scopes trial.

CHAPTER FIVE

How Scopes Was Framed

"I invoke Chemistry as an argument against Evolution," wrote William Jennings Bryan three days before Christmas 1923. "I believe that it annihilates the whole hypothesis of evolution. If Chemistry does not reveal a progressive force, such as evolutionists claim *it is not there*."[1]

Bryan was a religious leader who, earlier in the year, was nearly elected moderator of the General Assembly of the Presbyterian Church in the U.S.A. He wrote a weekly column of Bible lessons that was syndicated in newspapers around the country. The thrice-nominated presidential candidate was probably the most famous antievolutionist of the early 1920s. Why would he raise *this* argument against evolution? If he were truly interested in defending the Bible, as he often claimed, why would he resort to chemistry?

The answer is that Bryan thought that evolution was unjustified from both a scientific and a religious perspective. Rather than seeing science and religion as conflicting, he thought that both provided insights into a common truth. He made this point in a 1922 *New York Times* editorial: "If it could be shown that man, instead of being made in the image of God, is a development of beasts we would have to accept it, regardless of its effect, for truth is truth and must prevail."[2] Bryan recognized that he would need a scientific argument to *disprove* the theory of evolution. But he demanded the same from evolutionists. If chemistry did not disprove evolution outright, it posed a challenge that (in his view) biologists had not explained.

Only after asserting that there was insufficient scientific proof either for or against evolution did Bryan turn his attention to a pragmatic consideration of religion: "When there is no proof we have a right to consider the effect of the acceptance of an unsupported hypothesis."[3] He cited a study showing that teaching evolution undermines faith in religion and in morality.[4] Since evolution was not a proven fact, he argued, the *effects* of teaching it should resolve the matter of what should be allowed in schools. For Bryan, religion was the basis of morality, and, unlike the creators of civic biology education, he regarded the idea that one could teach morality apart from religion as absurd. Bryan's *New York Times* piece was one of the first public statements to attack evolution on these grounds, linking the religious claim that evolution did not happen with the moral effects of presenting the theory to students. This article may be considered a beginning of the school antievolution movement.

From the onset, school antievolutionism was not conceived in a spirit of science-religion conflict. Though conceiving science and religion as distinct enterprises was most memorably associated with those who did so in order to claim inherent conflict between the two, many who discussed the idea of a relationship between science and religion emphasized harmony. In fact, antievolutionists *and* their opponents frequently asserted that there could be no true conflict of "science and religion." Each side did accuse the other of fomenting conflict by misrepresenting either the nature of science or the nature of religion. While this debate took place within what Jon Roberts has called "the trope 'science and religion,'" partisans on both sides of the school evolution issue had different understandings of the components of that trope: *science* and *religion*.[5]

Considering the way in which biology education was brought to places like the rural South and the ideological agenda associated with biology textbook authors, it is easy to understand the emergence of the school antievolution movement without appealing to a conflict with the biblical account of creation. Religion initially intersected the school antievolution movement, not because evolution disproved Genesis, but because the issue of compulsory public schooling was frequently seen in religious terms, especially in the South. In the same legislative session that passed the antievolution law, a bill was introduced in the Tennessee House "to make it unlawful to employ atheists as teachers."[6] Control of the schools was viewed by many Tennessee parents as a matter of cultural and religious identity. In her history of rural Tennessee life, Jeanette Keith explained the connection between religious values and school reform in the Cumberland valley, which was represented by the author of the antievolution bill, John Washington Butler: "To reform the culture, as progressives tried to do in the early twentieth century, would mean a symbolic rejection of the ways of the fathers."[7] As Charles A. Israel has shown, much support for education in Tennessee had roots in religious values. Many of the most prominent advocates of public schools in Tennessee were motivated by their religion.[8] Although the antiatheist bill did not become law, its introduction was another example of how commonplace religious language was in school legislation in the state. The antievolution bill may still have been exceptional, however, as it directly invoked the Bible, something the antiatheist bill did not do.

Butler explained his reasons for proposing the antievolution law in an interview conducted two years after the Scopes trial:

> I had observed that parents were sending their children away from home to schools and at the time they were sent to school they respected the religion and the Bible believed in by their parents. When they went to college for a time, in many cases they returned home without any respect for the faith of their Christian fathers and mothers, thereby breaking the hearts of their parents, by telling them that the Bible is not true, that no one believes the Bible except ignorant and uneducated people; that the story of creation, as taught in the Bible, is only a fairy tale; that man came here not as the book of Genesis says, but by a process of evolution.[9]

Butler's reasons echo the pragmatic argument made by William Jennings Bryan: that teaching students "that man descended from a lower order of animals" had the *effect* of turning them away from religion and morality.[10] They also speak to concerns about education and cultural identity, evoking the prejudice that people who believe the Bible are uneducated. In an interview conducted during the trial, Butler said that he had "read in the papers that boys and girls were coming home from school and telling their fathers and mothers that the Bible was all nonsense."[11] It makes more sense that he *read* about this phenomenon rather than that he "observed" it (as he said in the later interview) because, at the time he proposed the law, there were almost no high schools in his legislative district.[12]

Butler's account of his own motivations clearly changed as a result of the Scopes trial, which he witnessed. He stated that he had never expected his law to come to trial.[13] Neither had Governor Austin Peay, who had described the bill as a "protest against an irreligious tendency to exalt so-called science."[14] As so often happened in Tennessee, religion motivated his political efforts, including those to expand public schooling through the general education bill.[15] Peay recognized the concern that compulsory schooling would threaten local religious values and culture, but he did not think that the antievolution bill placed religion and science in conflict with one another.

But Peay may have been as wrong about this as he was in predicting that the law would never be applied. The tension between religion and science was written into the law's ambiguous text:

> AN ACT prohibiting the teaching of the Evolution Theory in all the Universities, Normals and all other public schools of Tennessee, which are supported in whole or in part by the public school funds of the State, and to provide penalties for the violations thereof.
>
> Section 1. *Be it enacted by the General Assembly of the State of Tennessee,* That it shall be unlawful for any teacher in any of the Universiti[e]s, Normals and all other public schools of the State which are supported in whole or in part by the public school funds of the State, to teach any theory that denies the story of the Divine Creation of man as taught in the Bible, and to teach instead that man has descended from a lower order of animals.[16]

As it was written, Butler's law seems to embrace antievolutionism as both a school control issue and a science-religion issue. The title describes the act as "prohibiting the teaching of the Evolution Theory" in schools "supported in whole or in part by the public school funds of the State." Without any reference to religion, that title suggests that the law is meant to protect against abuses of *state* control of schools. It even implies that a public school that was only funded locally (and not by the state) could be exempt from the law. The text of the act's first section, however, does not mention evolution at all, instead prohibiting the teaching of "any theory that denies the story of the Divine Creation of man as taught in the Bible, and to teach instead that man has descended from a lower order of animals." The implication is that the biblical account of the creation of man is incompatible with descent from a lower order of animals. However, the law could be interpreted as banning only accounts of human origins that *both* deny the truth of the Bible and endorse evolution. In the latter reading, which Scopes's attorneys would advocate, theistic evolution would be acceptable since it endorses evolution but does not deny the Bible. By this reasoning, the law could be applied against those who taught evolution only if they also explicitly denied the truth of the Bible.[17]

Tennessee's law came after antievolution amendments that had been attached to school legislation in other states. Oklahoma passed an antievolution provision in 1923 as an amendment to a bill that would provide free textbooks to the state's students. The amendment prohibited the adoption of textbooks containing evolution. The Oklahoma governor who signed the bill wanted the free textbook law more than he opposed the antievolution provision.[18] The Tennessee general education bill did not directly regulate textbooks; rather, it expanded the presence of schools and teachers in the state. When Butler's bill is seen as a response to the general education bill, it makes sense that it too did not focus directly on textbooks.

Instead, it made it "unlawful for any teacher . . . to teach" an account of the origin of humanity that denied the truth of the Bible. Even though textbook content determined (or was thought to determine) much of what teachers actually taught, the law's prohibition of the actions of teachers rather than the adoption of certain textbooks made it possible for violations to lead to a *criminal* case.

Perhaps Butler and Peay would both have been right in thinking that no one would run afoul of the law had it not been the case that several parties took advantage of the statute in order to ensure that a trial took place. The ACLU (a young organization at the time, with little presence outside the New York area) saw it as an opportunity to expand its fight for teachers' rights and ran newspaper advertisements offering to provide legal defense for a teacher who violated the law. The people of Dayton, Tennessee, concocted the idea of a test case in response to the ACLU's ad, believing that the public attention would aid their attempt to revitalize their economically stagnant region. The trial garnered national attention and became a spectacle by attracting the involvement of the celebrity lawyers William Jennings Bryan and Clarence Darrow, both of whom saw the case as an opportunity to make political points in front of a captivated audience. The Scopes trial took place because several of its participants connived to ensure that it would happen. These different groups each saw the trial as a useful way to achieve goals that had little to do with the guilt or innocence of John Scopes. They had very different goals, and the only thing they all agreed on was that a trial should take place. But there was no consensus regarding *why* a trial should take place, how it should be conducted, or what was at stake. In effect, the trial consisted of these different parties fighting over the meaning of the thing they had together created.

Peay's statement on signing the Butler Act was only the first of many efforts to frame perception of what the antievolution law meant—both legally and politically. Those efforts continued throughout Scopes's indictment, his trial, and its aftermath. They even continue in the twenty-first century as reinterpretations of what the Scopes trial meant, why it occurred, and what impact it has had shape contemporary debates about American education policy and the role of religion in public life.

CREATING THE "SCOPES TRIAL"

Scopes's memoir relates how he came to be indicted for violating Tennessee's antievolution law. He filled in for the regular biology teacher at Rhea

County Central High School in Dayton sometime after the March 21 enactment of the antievolution law. In May, the ACLU announced that it would defend any teacher who would agree to be prosecuted under the antievolution law. A group of men in Dayton saw this news in the *Chattanooga Times* and conceived the idea of a test case. None of the men involved had strong feelings about the antievolution law itself. None of them had ever been outspoken on matters of science and religion. But they saw an opportunity to attract national attention that could bring publicity and economic growth to Dayton. They asked Scopes if he would be willing to stand trial, and he agreed.[19]

From so simple a beginning, a small-town misdemeanor trial that never would have taken place without the consent of the defendant evolved into an event depicted as the greatest legal clash between science and religion since the trials of Socrates and Galileo. This depiction was largely by the design of the trial's creators. The people of Dayton sought to portray their town as an exemplar of the American way of life and invoked the epic theme of "science and religion" conflict to bolster their town as the appropriate venue for such weighty issues. The ACLU sought to advance the cause of academic freedom and the rights of teachers. William Jennings Bryan saw the case as an opportunity to spread his antievolution crusade.[20] In his memoir, Clarence Darrow stated that he volunteered his services to stem the influence of "fundamentalists" who believe that the Bible "was virtually written by the Almighty and is in every part literally true."[21]

Some small textbook publishers even saw the trial as an opportunity to gain ground on their competitors by quickly developing books appeasing antievolutionists. Most larger publishers saw the trial as potentially negative publicity. Editors at both Ginn and the American Book Company (ABC) implored their biology textbook authors (Benjamin Gruenberg and George W. Hunter) to keep away from the trial and say nothing to the media.[22] Despite their concerns, these publishers still benefited from the Tennessee legislature's decision to prohibit *teaching* evolution instead of regulating textbook content. The more the trial focused on science and religion instead of education policy, and the more it focused on evolution's relationship to the Bible than on biology textbooks, the more easily they could present their books as acceptable no matter how the trial turned out. The less focus the trial placed on what John Scopes specifically did, the better.

While Gruenberg and Hunter were both warned not to get involved, plenty of other people flocked to Dayton. As Edward J. Larson describes, many of the residents of Dayton left town, renting out their homes to trial participants, journalists, and others who traveled to be part of the audience.

The trial attracted "hucksters and proselytizers" who formed a carnival of evolution and religion. WGN transmitted audio from the trial through dedicated telephone lines to its broadcast towers in Chicago, and over two hundred reporters traveled to the small town.[23] A *New York Times* article covering the first day of the trial summed up both the crowd and the public perception of the event with the headline: "Cranks and Freaks Flock to Dayton."[24]

This was the occasion for the spectacle of the Scopes trial. As Michael Lienesch rightly notes: "One of the reasons the Scopes trial was so significant was precisely because it was a good show."[25] The promotion of Dayton, the castigation of Fundamentalism or evolutionism, the great popular selling of ideas, and the battle for American public opinion had little need for the actual court case. As one article described it, the case was about "evolution and freedom of teaching" against "Christianity" and "the acceptance of the Bible as the literal word of God." Promising an event steeped in these issues, it proclaimed: "These will be the things which will be discussed when this case opens, not whether Mr. Scopes defied a State law and taught evolution. That is a minor issue."[26] The "Scopes trial"—as opposed to the court case *Tennessee v. John Scopes*—was going to be the "good show."

What made the Scopes trial a spectacle was not the court case that inaugurated it but the nature of its audience. It did not gaze passively at court proceedings unable to affect them; the audience created the public event by its very presence. The Scopes trial included the thousands of people who traveled to Dayton, reporters from all across the world, and their millions of readers and listeners. It extended beyond the courthouse, beyond the town, even beyond the United States. Though the court case *Tennessee v. Scopes* took place in Dayton, the Scopes trial took place across the globe; it was an event unfolding on the pages of newspapers and broadcast by the nascent medium of AM radio.

Not only was the spatial boundary between court and spectators blurred, but the separation between participants' roles as creators and consumers also became increasingly untenable. The jury was as neglected as the question of Scopes's innocence. In fact, the jurors heard less of the trial than anyone as they were often removed from the courtroom while lawyers argued over procedures. Not only did people involved in the trial react to the behavior of those consuming it from a distance, but their actions were also shaped by their *anticipation* of the audience. In public speeches, they reflected on how people would understand these events, and they were ever aware of the public gaze. There was a carnival atmosphere outside the court, with religious pamphleteers, scientific demonstrations, and people selling souvenirs. This self-awareness was not limited to the events that took place outside the legal case. The attorneys made reference to the courthouse crowd in their arguments and also evoked those paying attention beyond the Tennessee hills. What they, like the journalists surrounding them, were doing was composing a description of the trial that served the worldview they wished to create.

Not everyone involved in creating the Scopes trial had the same vision of what that worldview was supposed to be. The ability of trial participants to make their vision prevail depended in part on how well they could compete with opposing viewpoints expressed. But even when they took opposing sides, the trial's creators agreed that their debate was epic in scale and universal in the kinds of issues it discussed. Where the two sides agreed, or where one side's vision of the trial went unchallenged, their preconceptions became part of the consensus view of what the trial was. As Mary Beth Swetnam Mathews describes in her study of media representations of American fundamentalism, in their coverage of the Scopes trial "reporters drew upon a common language already used to describe the South and the fundamentalists."[27] Mathews particularly notes the presence of rhetoric associating the antievolution movement with the Inquisition and the Middle Ages, predisposing readers to reject it as outdated. Constance Clark writes: "Journalists crafted stories of the trial in the shape of allegory and the language of melodrama." But not everyone accepted these common tropes and ways of interpreting the evolution issue. In her study of the visual representations of evolution, Clark shows that the scientists who created images drew on "an established set of visual conventions" that other viewers did not always share.[28] What the scientists took as conventional others interpreted differently. The same was true of the language used to describe the Scopes trial.

Susan Harding points out that Fundamentalism and the idea of the "modern," especially in coverage of the Scopes trial, were often introduced in "binary opposition." She argues that setting concepts in such a binary opposition is itself a modern way of thinking and, therefore, that self-proclaimed Fundamentalists are also "constituted by modern discursive practices." Consequently, some Fundamentalists have effectively written their own histories using a narrative framework antithetical to the way of thinking they espouse.[29] This analysis is certainly true of Fundamentalist interpretations of the Scopes trial, its causes and its effects. Antievolutionist depictions of the trial as a battle between science and religion play into the modernist depiction of opposition that Bryan himself would have rejected.[30] The narrative of opposition was reinforced by the fact that this event began as a court case. The genre of a trial is inherently oppositional. One side prevails; the other does not. In a criminal case such as *Tennessee v. Scopes,* ultimately the jury must render a verdict of innocent or guilty—even though, in the Scopes trial, this outcome was never actually contested.

Scopes and his lawyers' legal reason for participating in the trial was to create a challenge to the Tennessee antievolution law in a higher court. In order to do this, Scopes could not plead guilty, nor could he be acquitted. The only people who had the opportunity to argue Scopes's innocence had no intention of doing so. Instead, his lawyers intended to show why the antievolution law was unjust even as they conceded Scopes's guilt in breaking it. It was the defendant's moral innocence—not his legal innocence—that they hoped to prove by arguing for their vision of the compatibility of science and religion. As the legal question of guilt or innocence was effectively ignored, the misdemeanor case of *Tennessee v. Scopes* was lost in the public debate about "science and religion" that was the Scopes trial. The attorney general wanted to keep the trial focused on the question of whether Scopes violated the law, not whether the law was just. But, just like Fundamentalists came to explain their participation in the Scopes trial in the binary mode put forward by their modernist opponents, the prosecution found itself engaged in a debate the other side had introduced. In no small part, this was because it had accepted the offer of assistance from William Jennings Bryan, who shared the interest in turning the Scopes trial into a spectacle. However, the impetus for turning the Scopes trial into a clash over science and religion came, not from Bryan, but from the defense attorneys: Darrow, Dudley Field Malone, and Arthur G. Hays. The defense succeeded in reframing the Scopes trial because, rather than contesting, it conceded the prosecution's legal position: *John Scopes was guilty.*

TENNESSEE V. SCOPES—GUILT BY TEXTBOOK

The prosecution's straightforward case that Scopes violated the antievolution law focused less on the teacher and more on the offending textbook. Instead of John Scopes, it was really Hunter's *Civic Biology* that was on trial. Rhea County school superintendent Walter White testified that Scopes had admitted to violating the law by using Hunter's textbook: "That theory of evolution; he said he couldn't teach the book without teaching that and he could not teach that without violating the statute." Two students from the biology class testified that Scopes taught them evolution. The prosecution had the two boys read from the *Civic Biology*, showing where it described the classification of man as a mammal.[31] The final prosecution witness was the school board member Frank Robinson, who owned the drug store where the *Civic Biology* was sold and where he and other Daytonians had convinced John Scopes to stand for trial. Robinson was also the coauthor of the pamphlet *Why Dayton of All Places?* which served to introduce Dayton to the world as a thriving and typical American town.[32] He testified that the *Civic Biology* was the book that was required by the state and that Scopes had admitted to using it.[33] For the prosecution, the case was simple. Scopes was guilty because he taught from Hunter's textbook. Teaching the book meant teaching evolution. Therefore, Scopes had violated the law.

The defense readily conceded that Scopes had used Hunter's textbook and that Hunter's textbook contained discussion of evolution. But its contention was that using the *Civic Biology* was not illegal because the law only prohibited teaching that "denies the story of the Divine Creation of man as taught in the Bible." The scientific theory of evolution, they claimed, did no such thing. Later in the trial, the defense returned to the matter of state-approved textbooks, attempting to show that the antievolution law did not conflict with teaching from accepted biology textbooks. Governor Austin Peay's signing statement, with his explanation that there was "nothing of consequence in the books now being taught in our schools with which this bill will interfere in the slightest manner" was not accepted as evidence, but the newly (June 1925) adopted biology textbook by Peabody and Hunt was admitted.[34] The defense used this to argue that state-approved textbooks taught the origins of man even after the law was passed.

In the opening statement for the defense, Malone introduced the question of the science-religion relationship, asserting: "Science occupies a field of learning separate and apart from the learning and theology which the clergy expound."[35] Because evolution and religion (as Malone defined them) are compatible, the law could not be applied.

This, in brief, was the crux of the defense strategy: to convince a court (the one that would hear the appeal of Scopes's conviction) that a reasonable interpretation of science and religion made the law vague and contradictory and, therefore, unconstitutional. That was really the primary legal basis on which the law could be challenged. While the American Federation of Teachers, the ACLU, and other organizations advocated for academic freedom, the rights of teachers in the classroom were not recognized by U.S. courts in 1925. No one made the argument that the law violated John Scopes's freedom of speech.

Freedom of religion was a more complicated claim. Antievolution trials in recent decades have focused on the question of the violation of the Establishment Clause of the U.S. Constitution,[36] but the U.S. Supreme Court did not rule that the religious freedoms guaranteed by the First Amendment applied to state laws until 1947.[37] However, the Tennessee state constitution also contains a "religious preference" clause that prohibits the state from taking any action that "gives preference to any religious establishment or mode of worship."[38]

So the defense's fight against the antievolution law could take one of two different approaches. First, it could argue that the law discriminated between two religious viewpoints, one evolutionary and the other antievolutionary, giving preference to the latter—in which case it would violate Tennessee's constitution (although other state antievolution laws would potentially be valid). Second, it could argue that espousing evolution and denying "the story of the Divine Creation of man as taught in the Bible" were not the same thing and, thus, that the law was unenforceably vague about what it actually prohibited. (The U.S. Supreme Court did not hold that a law could be invalidated for vagueness [as a violation of the due process clause of the Fourteenth Amendment] until a year after the Scopes trial, but the argument was already making its way through the court system.)[39] For either approach to succeed, the defense would have needed to show that some interpretations of evolution did not conflict with some interpretations of the Bible, that accepting evolution and denial of the Bible were not equivalent, and that there was a religion that accepted evolution—theistic evolution—and that was, therefore, injured by the state's preferential treatment of Fundamentalism.[40] The defense had to concede Scopes's actions and the contents of Hunter's *Civic Biology* and put forward its own interpretation of science and religion in order to make its case.[41]

THE SCOPES TRIAL AND SCIENCE-RELIGION CONFLICT

The presentation of Tennessee's antievolution law as a matter of science and religion began even before John Scopes was involved. At the end of March, just after the law was signed, Shailer Matthews, dean of the University of Chicago Divinity School (and later one of the expert witnesses for Scopes's defense), called the Tennessee law "one of the most dangerous pieces of legislation ever passed." A newspaper account of his speech on the law attributes to its supporters a belief "that science and religion are incompatible." Matthews, the article continued, "has been at pains to show it is not in the true nature of either" to conflict.[42] On the Sunday before Scopes was indicted, the World Christian Fundamentals Association, an organization of Protestant Fundamentalists of many denominations, held its annual meeting in Memphis. The program featured the talks "Evolution," by the association's founder, William Bell Riley, and "Evolution, or a Universe without God," by William Jennings Bryan.[43] It was at Riley's urging that Bryan decided to go to Dayton.[44]

Antievolutionists and their opponents had long invoked the question of science-religion relationships in describing their debate. But partisans on both sides said the same thing: that their own side held that science and religion were compatible while the other side mistakenly took them to be in conflict. This continued in Dayton. The two sides in the Scopes trial each had a different view of what counted as "science" and what counted as "religion" even as they each claimed that science and religion, considered together, did not conflict. Their legal debate became a conflict over whether theistic evolution and biblical literalism were viewpoints that belonged under the category *religion* and whether evolution deserved to be called *science*.

The defense came to be seen as proponents of biological science and opponents of faith in the truth of the Bible. The prosecutors came to be seen as defending scripture and opposing evolution. As a result, the conflict over the boundaries of the concepts called "science" and "religion" (within an overarching belief in their ultimate harmony) became a conflict *between* science and religion.

FROM ACADEMIC FREEDOM TO SCIENCE VERSUS RELIGION

The ACLU had viewed the antievolution law as part of a larger movement to reduce the rights of teachers. Laws requiring loyalty oaths, banning atheists from the classroom, and curtailing union organizing in schools had been passed by legislatures around the country. Combating antievolutionism was not the raison d'être of the ACLU, though with Henry Linville's involvement it was a cause that found strong support. The ACLU had not been involved with fighting antievolutionism before Butler's bill was enacted, but the Tennessee law was the first antievolution law to restrict teaching, instead of imposing conditions on what textbooks the state could use. That difference made the Tennessee law more clearly an infringement on teachers' rights.[45]

But claiming that the antievolution law was unconstitutional *because* it infringed on teachers' rights was a legally tenuous claim. As William Jennings Bryan argued: "The hand that writes the teacher's pay check is the hand that rules the schools."[46] This was an expression of Bryan's majoritarianism, the proposition that government should advance the view of the majority of the people.[47] Lawrence Levine suggests that, "to Bryan, the 'majority' meant the majority of those who came from those sections of the country which he represented and who fundamentally accepted the values he felt were essential."[48] In this view of majoritarianism could be heard overtones of support for states' rights, a position that had plenty of influence in the American courts, especially those in the South, at the time. Any attempt to challenge that principle with only a vague assertion of academic freedom had little hope for success.

Arthur G. Hays, the ACLU's general counsel, suggested that a better approach would be to focus on the issue of religious freedom. By claiming that there was a single understanding of Genesis—"*the* story of the Divine Creation"—the Tennessee law implied approval of some religious beliefs and intolerance of others. By discriminating against more modern religious groups that had incorporated notions of the evolutionary origins of humanity into their theology, the law amounted to an authorization of a particular religion and a violation of Tennessee's constitution. An appeal based on these grounds would have to go to the state supreme court, despite the hope of some in the ACLU (who distrusted southern state courts) to go directly to a U.S. federal court.[49] Before there could be any appeal, however, Scopes would have to be convicted in Dayton.

Once they started planning for the trial, the defense lawyers emphasized the antievolution law's conflict with religious freedom as opposed to academic freedom. To do this, it was important that evolution be portrayed as compatible with religion. Worldviews that took account of evolutionary origins of the human species had to be included in the domain of what was protected as religion. After several days of jury selection and hearings on motions, the trial began in earnest on Wednesday, July 15. After the prosecution made its case with its four witnesses, Malone laid out the defense's position:

> The defense believes that there is a direct conflict between the theory of evolution and the theories of creation as set forth in the Book of Genesis. . . .
>
> But we shall make it perfectly clear that though this is the view of the defense we shall show by the testimony of men learned in science and theology

> that there are millions of people who believe in evolution and in the stories of creation as set forth in the Bible and who find no conflict between the two. The defense maintains that this is a matter of faith and interpretation which each individual must determine for himself.[50]

To establish this, the defense attempted to call expert witnesses who would testify about the nature of evolution, its acceptance among scientists, and its compatibility with religion. By showing that these men find no conflict between science and religion, it hoped to show that the anti-evolution law discriminated against certain types of religion. The defense called Maynard Metcalf, a zoologist from Oberlin College and Johns Hopkins University, to the stand. Metcalf discussed his scientific training and his experience teaching Bible classes at his church (a Congregationalist one, though before that he had also been a member of a Presbyterian church).[51] The prosecution objected as soon as questions turned from Metcalf's background and qualifications to the subject at hand. Each time Metcalf was asked a question about evolution, prosecutors disputed the relevance of his testimony to the legal question of Scopes's guilt. Although he was permitted to give answers without the jury present (for the sake of creating a complete transcript), none of his testimony was allowed in the trial. The prosecution then objected to even calling the rest of the defense witnesses, claiming that none of them could speak to the question of whether Scopes had violated the law. As the argument over these witnesses grew more heated, the defense's portrayal of science and religion evolved.

On Thursday, July 16, the defense tried again to bring expert witnesses to the stand. Darrow reiterated their role in the defense theory of the case:

> We intend to show by men of science and learning, both scientists and real scholars of the Bible, men who know what they are talking about, who have made some investigation, expect to show, *first*, what evolution is, and, *secondly*, that any interpretation of the Bible that intelligent men could possibly make is not in conflict with any story of creation, while the Bible, in many ways, is in conflict with every known science, and there isn't a human being on earth who believes it literally.
>
> We expect to show that it is not in conflict with the theory of evolution.
>
> We expect to show what evolution is, and the interpretation of the Bible that prevails with men of intelligence, who have studied it.[52]

As Darrow framed it, science and religion were not in conflict, but religion was not the same thing as the Bible itself. Religion was what people *believed* about the Bible. Echoing Malone's earlier claim that religions must be subject to individual interpretation, Darrow argued that no "men of intelligence" would—or could—interpret the Bible *literally*. Darrow and the defense were in effect claiming that the antievolutionist position, as they understood it, was not a religious position at all. Religion in their view was a human act of belief and intellection. The defense strategy was to argue for the human interpretation of the Bible, the legitimacy of evolution as a scientific theory, and the subsequent compatibility of evolution and the Bible within that framework.

SCIENCE AND RELIGION: FROM HARMONY TO CONFLICT

The defense's plan to promote a vision of science and religion as harmonious was thwarted by the judge's ruling on Friday, July 17, that its expert witnesses were irrelevant to the legal question of Scopes's guilt. However, another tactic was made available after William Jennings Bryan spoke on Thursday and engaged the defense's issue of the proper relationship of science and religion (see fig. 3). John Scopes recounted: "The technical point being argued was forgotten and was not, to my knowledge mentioned after the first five minutes of Bryan's speech. Bryan addressed the Judge and then immediately turned to face the spectators. There was no pretense; this was to be a speech to the people, not merely to the court."[53]

Bryan began by reiterating the prosecution's case that the experts were irrelevant and that all that mattered was that Scopes had taught evolution from Hunter's *Civic Biology*. The textbook became the centerpiece of Bryan's attack. In that book, he explained, was a clear indication of what was wrong with teaching the theory of evolution:

> There is that book; there is the book they were teaching your children, that man was a mammal and so indistinguishable among the mammals that they leave him there with 3499 other mammals, including elephants! Talk about putting Daniel in the lion's den. How dare those scientists put man in a little ring like that with lions and tigers and everything that is bad! . . .
>
> [Hunter] tells the children to copy this, copy this diagram! In the notebook children are to copy this diagram and take it home in their notebooks, to show their parents that you cannot find man! That is the great game to put in the public schools, to find man among the mammals, if you can![54]

Bryan did not limit himself to the classification of man and mammals, however. He moved quickly to reiterating arguments he had often used in other

FIGURE 3 William Jennings Bryan speaking at the Scopes trial on Thursday, July 16, 1925. Bryan addressed a packed courtroom and the audience listening to WGN radio as well as the court itself. Courtesy of Tennessee State Library and Archives.

speeches: that teaching evolution eroded the moral sense of students; that missing links and examples of one species developing from another had not been found; and that evolution contradicted the Bible.

Just a month before the trial, Bryan had told a colleague: "Hunter's Civic Biology . . . certainly gives us all the ammunition we need."[55] But, realizing that the defense's plan was to concede the *legal* guilt and claim that Scopes was *morally* innocent of any wrongdoing, Bryan reached for his arguments against the morality of teaching evolution. Evolution undermined morality, not because it disputed the biblical account of human origins, but because it denied the Bible's promise of redemption. By presupposing that all change in the world was gradual and explainable by natural law, evolutionists ruled out even the possibility of miracles. But Bryan did not claim that evolution contradicted the Bible because it denied the specific age of the world or the duration of the days of creation. Even if scientists "admitted that God created the cell," Bryan proclaimed, "they did not tell us where in this long period of time between the cell at the bottom of the sea, and man, where man became endowed with the hope of immortality."[56] It was William Jennings Bryan who opened the door for the defense to turn the prosecution's narrowly constructed case concerning what John Scopes had taught out of George W. Hunter's textbook into a debate over science and religion.

SCIENCE, RELIGION, AND BIBLICAL "LITERALISM"

As recent historians of the Scopes trial have noted, Bryan did not believe in doctrines typically associated with a literal interpretation of the Bible. But he did endorse a different kind of literalism. For him—as for most Fundamentalist Protestants in the early 1920s—"literalism" referred to the *authorship* of the Bible, to the fact that every word of the Bible was as God inspired. To the letter, the Bible was right and true. But this belief did not endorse the literal *interpretation* of the Bible. Many literalists, including Bryan, held that the human act of interpretation could not provide the same kind of certainty as that revealed by God. Bryan expressed doubt about the interpretation that the earth was only a few thousand years old and that the six days of creation were each twenty-four hours long: "I think it would be just as easy for the kind of God we believe in to make the earth in six days as in six years, or in six million years, or in six hundred million years. I do not think it important whether we believe one or the other."[57] It was not a devotion to a young earth that led Bryan to claim that evolution was incompatible with the Bible.

Bryan emphasized that in an evolutionary account of origins there is no room for miracles, the soul's immortality, the virgin birth, the resurrection of Christ, or salvation. The belief in a world beyond what people could understand as natural was the essence of what Bryan took to be religion. To rule out the possibility of supernatural causes was inherently atheistic. To make theology contingent on scientific discovery meant that revelation or inspiration found in a literally true Bible had no place. Theistic evolution was a logical impossibility for Bryan. It suggested that scientific discoveries were true no matter what insights divine inspiration might provide. It rejected the possibility that the Bible was literally true because it ruled out divine authorship. But Bryan did not claim that evolution was wrong because it conflicted on points of fact with a literal *interpretation* of the scriptures. It conflicted with the belief that there were truths beyond science and human reason. There could not be an evolutionary religion. Evolution, as a scientific methodology, undermined the very basis of religion. The defense arguments—that the antievolution law gave preference to antievolutionary religions over evolutionary religions or that the law was vague because evolution did not contract the story of creation in the Bible in an evolutionary religion—were faulty because the very idea of an evolutionary religion was incoherent.

Bryan was drawn into the science-religion framework that Malone and Darrow had put forward. He replied to their constitutional arguments against the antievolution law but did so in a way that allowed the legal prosecution of Scopes to be overwhelmed by a debate over whether science and religion were reconcilable. For Bryan, theistic evolution was not just incoherent; it was also dangerous. "Theistic Evolution," he wrote in his 1922 *In His Image*, "may be defined as an anesthetic which deadens the patient's pain while atheism removes his religion."[58] Scopes's defenders were advocating a reinterpretation of what religion itself was that would ultimately destroy faith. Regardless of the legal outcome, the use of the trial to popularize this view of religion was the real threat. "They did not come here to try this case," Bryan would declare later in the trial. "They came here to try revealed religion. I am here to defend it."[59]

Almost as soon as Bryan finished his speech that Thursday, Malone rose for the defense. He pointed out that the public attention paid to the trial proved that it was about bigger issues than a misdemeanor. He also made it clear that the prosecution, in the person of William Jennings Bryan, could no longer claim that the case was merely about Scopes's violation of the law: "What is the issue that has gained the attention not only of the American people, but people everywhere? Is it a mere technical question of whether the defendant Scopes taught a paragraph in the book of science? You think, your Honor, that the News Association in London . . . is interested in this case, because the issue is whether John Scopes taught a couple of paragraphs out of this book? Oh, no, the issue is as broad as Mr. Bryan himself has made it."[60]

Bryan's speech and Malone's response made it nearly impossible to return to the question of what Scopes had taught. The defense had succeeded in making the case a trial of the science-religion relationship. But with Bryan's entry into that discussion the defense changed tactics—from assuring that religion (in the guise of theistic evolution) and science were compatible to emphasizing the total rejection of biblical literalism and the so-called religion of William Jennings Bryan.

For Clarence Darrow and the other defense attorneys, the literal interpretation of the Bible was not really a position that could be seen as religious. As he had said shortly before Bryan's speech: "Any interpretation of the Bible that intelligent men could possibly make is not in conflict with any story of creation, while the Bible, in many ways, is in conflict with every known science, and there isn't a human being on earth who believes it literally." Like others who had attacked Bryan's antievolutionism over the years, Darrow attributed to Bryan a belief in the literal interpretation of the Bible. Both Bryan and Darrow agreed that the primary question in the Scopes case was that of how to understand the science-religion relationship, but each saw the other's religious position as not a part of "religion" at all.

This came to a climax on Monday, July 20. With trial proceedings moved to the grounds outside the courthouse (in response to concerns for the safety of the audience inside a hot and overcrowded structure), the defense asked for a sign stating "Read Your Bible" to be removed from the sight of the jury, lest it unduly influence them.[61] In response to the attorney general's suggestion that both sides should want to encourage Bible reading, Darrow offered this argument: "Mr. Bryan says that the Bible and evolution conflict. Well, I do not know, I am for evolution, anyway. We might agree to get up a sign of equal size on the other side and in the same position, reading, 'Hunter's Biology', or 'Read your evolution.'"[62]

In this reply, Darrow seemed to accept the view that he had attributed to Bryan, "that the Bible and evolution conflict," and align himself with the evolution side. He also aligned evolution with Hunter's *Civic Biology*, holding it as against the Bible and against the law. Darrow had re-presented a debate over different interpretations of *how* science and religion were harmonious as a debate *between* science and religion. This was part of the plan he had devised in response to the prosecution's victory the previous Friday when the judge excluded the defense's witnesses. The judge ruled that they were irrelevant to the question of Scopes's guilt. However, he permitted the defense to read prepared statements by the experts into the record of the trial without the jury present (so that an appellate court could read them). It began when Arthur G. Hays announced that the defense would like to call William Jennings Bryan to testify: "The defense desires to call Mr. Bryan as a witness, and of course, the only question here is whether Mr. Scopes taught what these children said he taught. We recognize that what Mr. Bryan says as a witness would not be very valuable. We think there are other questions involved, and we should want to take Mr. Bryan's testimony for the purposes of our record—even if your Honor thinks it is not admissible in general—we wish to call him now."[63]

What followed was one of the most famous scenes in American legal history. Darrow attacked the literal interpretation of the Bible, ridiculing the possibility of events like Joshua making the sun stand still and Jonah being swallowed by a whale. He mocked as unscientific the ideas that the world was only six thousand years old and that there had been a worldwide flood. Bryan defended the truth of the Bible but also argued that he was not trying to *interpret* the text.[64] He did not defend the literal interpretations that Darrow put forward, but at every turn he maintained that the Bible was true. He defended the literal *inspiration* of the Bible. Darrow seemed not to take note of the distinction and continued to use Bryan's appearance on the stand to mock the "fool ideas that no intelligent Christian on earth believes."[65]

Historians disagree as to whether this exchange was a resounding victory for Darrow, whether Bryan defended his religious views successfully, or whether Darrow came off as bullying.[66] Some commentators have even suggested that Bryan's death one week after the trial's conclusion was the result of his disappointment at the outcome of (or his overexertion at) the trial. But all agree that it was the culmination of a spectacle that had been building since the trial began and that it was the debate that proved to the world the reality of the conflict between religion and science.

VERDICTS ON THE SCOPES TRIAL

After the Bryan-Darrow exchange, the trial quickly drew to a close. The next day, the defense asked the jury to find their client guilty so that it could appeal the constitutionality of the antievolution law.[67] This was done, and the court proceedings ended. But the jury's decision had already become a triviality. In the most banal sense, the outcome of the Scopes trial was a foregone conclusion. But no one at the start of the trial could have anticipated some of its other effects. The Scopes trial helped recast popular understandings of "science and religion," biblical "literalism," and the "antievolution" law. This was due to the fact that it was never really a *trial*; that is, it was never an attempt to assess evidence and argue the law so as to determine a matter of guilt or innocence. Scopes's guilt was a given. As a result, the arguments that led to his conviction must have been valid arguments. Science and religion must really be in conflict. Antievolutionism must mean an affirmation of biblical literalism, and literalism must mean an act of interpretation. The Tennessee law that Scopes violated must prohibit what he did, which means that whatever was in Hunter's *Civic Biology* was illegal.

The idea of science-religion conflict was hardly new. It had roots in the late nineteenth century with the books of both John Draper and Andrew

Dickson White. But the Scopes trial aligned the science-religion conflict with the social conflicts that had given rise to the school antievolution movement. In doing so, that conflict ceased to be about just the incompatibility between different ideas and became the basis for competition between different groups of people. It is also very likely that the trial's association of antievolutionism with the literal *interpretation* of the Bible contributed to the increased number of self-described literalists who insisted on a young earth interpretation of Genesis after 1925. (Very few creationists espoused the notion of a young earth prior to the trial, but that viewpoint has since become widely embraced among the American population.)[68]

In addition to denoting a relationship between two separate sets of ideas, communities, or ways of interpreting the world, the phrase *science and religion* became something more than just *science* and *religion*. If the trial had been presented as just a matter of religion, few would have found it more memorable than the revival meetings that sprung up around the country every year. If it had been depicted as a trial of science alone, it might have attracted little more public notice than an academic conference. But the *world* came to Dayton. News reports were sent daily across the Atlantic. The trial inspired drama, songs, and art and maintains a prominent place in U.S. history. The reason for this intrigue was neither science nor religion but the spectacle associated with "science and religion." This re-presentation of the trial was exactly what most of the trial's participants wanted. It also served to obscure many of the issues that had actually contributed to the trial.

Austin Peay was correct when he wrote that there was "nothing of consequence in the books now being taught in our schools with which this bill will interfere in the slightest manner." But his conclusion that, "probably, the law will never be applied" could not have been more wrong.[69] The metamorphosis of *Tennessee v. Scopes* into the Scopes trial happened because the answer to the legal question of guilt or innocence was a foregone conclusion, allowing the philosophical issues that warranted the spectacle to proceed unabated. None of this would have been possible without the undisputed assumption that John Scopes was guilty of violating the antievolution law. The defense lawyers told the jury to convict him; Scopes himself admitted he was guilty. The legal case was trivial, its outcome assured.

SCOPES'S APPEAL

The greatest irony of the whole trial is that, arguably, Scopes was *innocent* of violating the Butler Act. In his memoir, Scopes wrote: "I didn't know, technically, whether I had violated the law or not." At least, not until he re-examined the *Civic Biology*. "There's our text, provided by the state. I don't see how a teacher can teach biology without teaching evolution."[70] The certainty of his guilt rested within the pages of the *Civic Biology*. Scopes saw that the textbook used the word *evolution*. Therefore, it violated the anti-evolution law. The prosecution had the school board superintendent and students examine the book, pointing to passages that showed man classified among the mammals. In his great speech during the trial, Bryan attacked the *Civic Biology*. Scopes was guilty. Scopes taught from the *Civic Biology*. Therefore the *Civic Biology* must have content that violates the law.

Except that it doesn't. The Tennessee law specifically prohibited teaching "that man has descended from a lower order of animals." Hunter's book talks about the *classification* of "man," arguing that he belongs among mammals—and specifically among primates—because of similarities of anatomy and physiology. The *Civic Biology* does not explicitly claim that classification implies common descent. The book talks about the development of modern humanity from "races much lower in their mental organization than the present inhabitants."[71] But it never says that humans descended from a *non-human* species. Unlike some other textbooks, Hunter's *Civic Biology* never talked about a "missing link," Cro Magnons, Neanderthals, Java Man, or any other human ancestors. (By contrast, Gruenberg's *Elementary Biology* included all of these.)[72] George Hunter certainly believed that humans shared ancestors with nonhuman species, but the *Civic Biology* never explicitly says that.

Constance Clark and others have noted that Bryan was particularly concerned with a diagram in Hunter's book labeled "the evolutionary tree." This diagram shows mammals as a small circle representing 3,500 species branching off from a trunk line that includes birds, reptiles, and other vertebrates, radiating along with other collections of species from a point near protozoa.[73] Despite what Bryan said in the courtroom—that man was "there with 3499 other mammals"—humans were not listed on this evolutionary tree (see fig. 4). Darrow even pointed this out when cross-examining the prosecution witness Frank Robinson.[74] As Clark remarked: "Visual images played an important part in the public discourse associated with the Scopes trial, but they did not necessarily convey the messages their authors intended." Hunter's explanation in the surrounding text could suggest to students that an evolutionary tree shows common descent, or it could merely indicate, as the text says, an "arrangement . . . called the *evolutionary series*" that shows the similarities by which animals are classified and the apparent

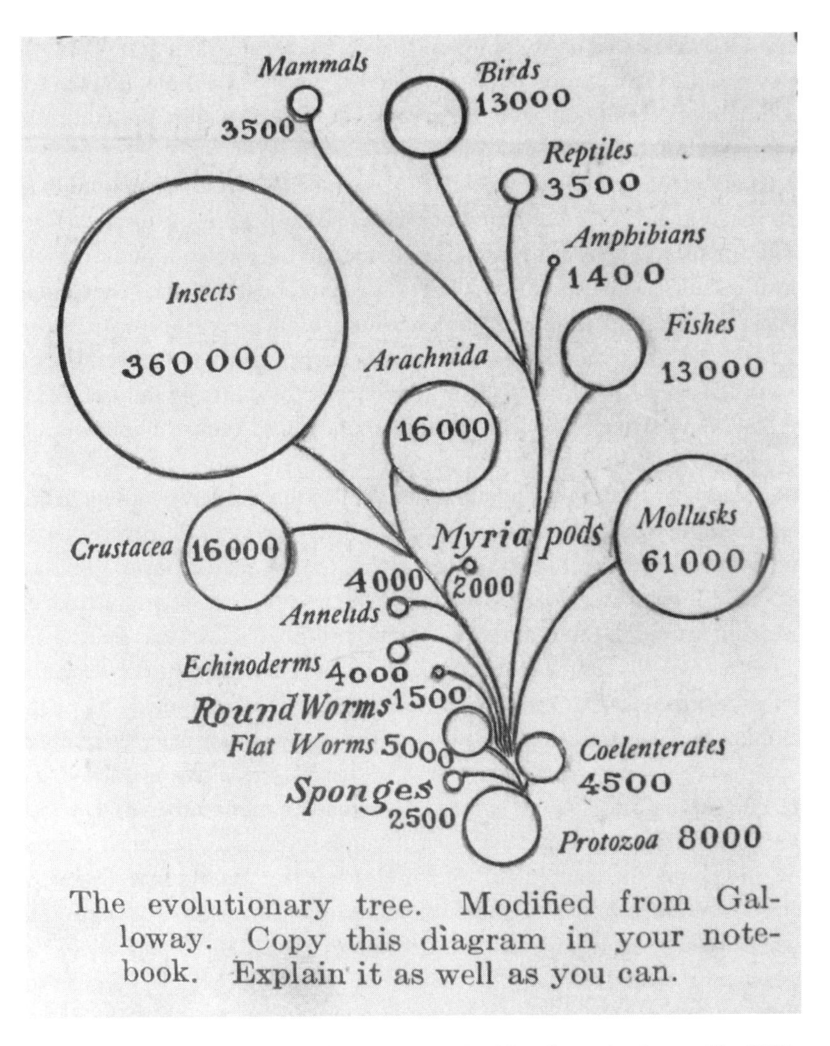

FIGURE 4 The "evolutionary tree" in Hunter's *Civic Biology* (p. 194) referenced by William Jennings Bryan in his attack on the textbook. Despite Bryan's interpretation, humans are not explicitly listed in the tree.

age of their emergence based on "stages in complexity of development of life on earth."[75] The widespread presentation of evolutionary trees—many of which explicitly claimed to represent ancestry and included humans—makes a compelling case for the "established set of visual conventions" that would lead a viewer like Bryan in 1925 to unambiguously interpret the diagram as inclusive of human beings and indicative of actual descent, not merely as a representation of classification.[76] But like the Bible, like the trial itself, Hunter's textbook gave rise to *interpretations* that were not controlled by its creators.

Though Hunter's text states that man must be *classified* with mammals for reasons of anatomical similarity, man does not appear in the *Civic Biology*'s evolutionary tree, nor is he explicitly claimed to have any nonhuman ancestor or relation.[77] The students called to the stand never testified that Scopes taught that humans evolved from a nonhuman ancestor; they testified that he reviewed the *Civic Biology* with them. Scopes could have been acquitted if his lawyers had wanted to argue that the *Civic Biology* did not violate the law. Scopes was innocent—if both the law and the textbook are read literally.

Someone must have been telling lies about John Scopes, he knew he had done nothing wrong, but, one morning, he was arrested. He agreed to submit to the test case and to suffer conviction for the sake of testing the law, but, had his guilt or innocence actually been in question, he might have deserved an acquittal. Instead, it was a given truth that Scopes was guilty—and guilty because he taught Hunter's *Civic Biology*. As a result, the textbook, already overdue for a new edition, lost sales even more quickly as people assumed that it was not compliant with the antievolution law. Virtually all media reports stated that Scopes was on trial for teaching evolution, not the nonhuman ancestry of humanity. Given the ease with which people looked at the textbook and saw that it taught evolution, the antievolution law, outlawing the teaching of human descent from nonhuman forebears, became an anti*evolution* law, making mention of any kind of evolution taboo. Darrow's suggestion that a sign stating "Read Hunter's Biology" or "Read your evolution" would balance out the objectionable "Read Your Bible" banner set the textbook up as the opposite of the Bible. Read as irreligious and in violation of the antievolution law, the *Civic Biology* was left with a stigma that the ABC would have to overcome. No textbook publisher, no state textbook commission, no school board, would risk splitting hairs over what the law taught.

CHAPTER SIX

The Evolution of the *New Civic Biology*

> The one name indelibly associated with the word evolution is that of Charles Darwin.
>
> *A Civic Biology* (1914)

> Another important line of biological investigation is the study of heredity and of the development of life on the earth. The name of Darwin is most indelibly associated with this branch of biology.
>
> *New Civic Biology* (1926)

"Have you heard anything about that Tennessee teacher's trial for teaching of Evolution? I heard they used a Hunter as evidence in this trial." "And what will be the result of the Schop [*sic*] trial in our sales? The book certainly received enough publicity out of the newspapers."[1] These two queries came on May 25 and June 2, 1925, before the Scopes trial even began. Both were hastily scribbled postscripts to poorly typed letters from George W. Hunter, the author of *Civic Biology*, to American Book Company (ABC) editor in chief George W. Benton. The second of these was written before any reply to the first could even have reached Hunter in La Jolla, California.

Early reports of John T. Scopes's arrest in the first days were conflicting. On May 7, newspapers around the United States carried word of the trial of, variously, "Scoaps" (*Columbus Journal*, *Chicago Tribune*, and *St. Louis Star*), "Scopez" (*Minneapolis Morning Tribune*), and "J. P. Scoats" (*Kingston [NY] New Leader*), who had been charged with teaching evolution in Tennessee. The *Los Angeles Herald* referred to both "Scoaps" and "Schoaps" *in the same article*.[2] As the variety of names suggests, early stories were full of factual errors, but there was a haste to cover an antievolution trial. By May 10, most accounts had Scopes's name right, but the *Chicago Tribune* reported that the offending textbook was "A Civic Biology" by "Leone W. Hunters" (an error most likely due to someone poorly handwriting the author's name in script).[3] It is not clear where George Hunter first read about the "Schop" trial or the fact that one of his books was involved, but his sense that the trial might mean good publicity for the *Civic Biology* betrays the kind of naïveté about the *business* of textbooks that often had ABC editors throwing up their hands in frustration with him.

The *Civic Biology* was not selling well in 1925. Sales had been declining for a few years as other publishers had civic-oriented biology textbooks that were newer. Hunter's textbook was eleven years old; the consensus among publishers was that the maximum life for a textbook was about ten years. To make matters worse, in 1923 the ABC published Hunter's *New Essentials of Biology*. This was intended as a ninth-grade biology textbook for rural schools. ABC salesmen reported that it was actually eating into sales of the *Civic Biology* but was of no use in helping compete against tenth-grade civic biology texts from other publishers. In the months before the Scopes trial, ABC sales agents were reporting trouble all over the country because they lacked a new enough biology textbook. "It is unfortunate," Benton wrote on April 23, "that so many of our battles have to be fought with guns not yet built."[4] Hunter was already working on a revision of the *Civic Biology* and was very near to finishing by the time news of Scopes's indictment reached him in California. On June 7, word came to Benton that the ABC had lost Detroit. The selection committee there had preferred a competitor's book over the *New Essentials*; however, adoption was postponed "on the assurance that [a new *Civic Biology*] would soon be ready."[5] But, with Hunter still at work on the manuscript, there was no possibility that a new edition of the *Civic Biology* would be ready in time for the start of the next year. Every week seemed to cost more sales, and to delay beyond this upcoming trial would certainly hurt business even more. Apparently, Hunter was also being asked to comment to the press once the textbook involved was correctly attributed to him. On June 8, Benton wrote to Hunter: "None of us know how far-reaching this controversy is likely to be. For that reason I urge you to refer all questions or invitations to talk on the subject to your publishers."[6] Two days later, Benton was more direct: "Whatever you do, please do nothing and say nothing."[7]

In their groundbreaking 1974 study, Judith V. Grabiner and Peter D. Miller claimed: "Hunter himself had been concerned that the Scopes trial publicity would drive his book out of the classroom, so he was willing to make changes."[8] They invoked a secondhand account from one of Hunter's colleagues, reported nearly half a century after the trial, to conclude that Hunter was concerned about the trial's negative publicity hurting his textbook. Perhaps Hunter himself revised his recollections to bring his own concerns into alignment with his editors', but his correspondence at the time suggests a very different story. He seemed to imagine that the publicity of the trial would help sales of his book in regions of the country where support for evolution was strong. It was for precisely that reason that he argued against altering the revised manuscript to accommodate antievolutionists in any way.

Historians of education have frequently raised the question of what effect the Scopes trial had on the content of biology textbooks. Starting with Grabiner and Miller, comparison between textbooks published before the trial and those published after has been a standard practice used to answer this question.[9] But it is not without shortcomings.

By 1925 the aging book was already out of most classrooms in the United States, and in mid-June it was also gone in Tennessee.[10] Looking at the *New Civic Biology* next to its 1914 precursor, it is easy to imagine that Hunter saw what had happened in the Scopes trial and hurried out a new book to avoid all the criticisms the trial had raised. But doing so assumes that the person described as the "author" of the textbook has much greater control over its content than was actually the case or that he and his publisher are in general agreement over what a book ought to say. At the end of their comparison between the *Civic Biology* and the *New Civic Biology*, Grabiner and Miller attribute all the key differences they found to post–Scopes trial appeasement *by Hunter*: "Should there be any doubt as to why these changes were made, the treatment of man's place in nature should indicate why; it begins with the newly added sentence, 'Man is the only creature that has moral and religious instincts.' These changes are relatively minor and presumably made at the last minute. But they indicate the attitudes of the textbook publishers and authors of the time."[11]

In his history of the Scopes trial, Edward J. Larson makes a similar assumption about authorship, claiming that Hunter was responsible for the addition of the "moral and religious instincts" line.[12] But this conclusion is unwarranted. The added sentence came from an ABC history textbook published in 1921.[13] It was included at the suggestion of a local sales agent in Alabama, whose recommendation was passed through the Cincinnati sales office to the New York editorial department in April 1926.[14] George Hunter had nothing to do with it. The people responsible for adding it did so for their own reasons. It is a mistake to attribute all the differences between the *Civic Biology* and the *New Civic Biology* to Hunter, just as it is a mistake to credit the Scopes trial with being the sole inspiration for them. Religious antievolutionism was not seen as a serious threat by ABC editors and salesmen when the *Civic Biology* first came out. The *New Essentials of Biology* was published in 1923, and, despite its rural focus, it included mention of evolution. But, by the end of that year, some textbook publishers had started to look for ways to respond to the school antievolution movement.[15]

The first revision of the *Civic Biology* that Hunter provided to the ABC (shortly before the Scopes trial began) was not nearly so sensitive to antievolutionism as its recently published competitors. Much of that sensitivity probably came from the rival publishers' editorial rooms rather than from Hunter's fellow authors, but it is clear that responses to antievolutionism predate the Scopes trial itself. Still, the Scopes trial changed what people thought *antievolutionism* meant. Even though the *Civic Biology* arguably conformed to the letter of Tennessee's law, Scopes's conviction, and the tacit vilification of the textbook that both sides of the trial took part in, made it impossible for anyone to claim that it could still be used in an antievolution state. The association of the *Civic Biology* with Scopes tainted the perception of the book, of George Hunter as its author, and the ABC as its publisher. The new book would have to be demonstrably different in order for its salesmen to claim that it was suitable for use in places that wanted to avoid controversy. This did not become obvious to the ABC's editors until quite some time after the trial ended, and Hunter never stopped insisting that making such changes would be exactly the wrong thing to do.

The development of the *New Civic Biology*, lasting from February 1924 to May 1926, was certainly affected by the Scopes trial, but the revisions were neither quick nor uncomplicated. The production of the revision included the input of the author, several editors, managing partners, sales agents, and a good number of people who were not even part of the ABC. As a book that was widely presented in the light of its predecessor's use in the Scopes trial, the *New Civic Biology* was uniquely positioned to help determine what would count as a response to the trial and how post-Scopes textbooks should be read. The creation of this response relied on feedback from teachers and regulators and the active role of textbook salesmen to promote certain interpretations of their books to users.

GEORGE HUNTER AND THE ABC

Hunter's relationship with the ABC had never been an easy one. His resistance to altering the *Civic Biology* to facilitate the 1915 Boston adoption took nearly six months to overcome, with Benton threatening to hold up publication of the laboratory manual for the textbook and Hunter threatening to sue in retaliation. Hunter eventually made the changes the editors requested, but the ABC was already beginning to look for other biology authors. ABC salesmen were quite happy with the new *Civic Biology* but suspected that the 1911 *Essentials of Biology* would not do for rural textbook markets. "I believe that Hunter's book was written from the standpoint of the city community, and that there is probably room for another book along similar lines written from the standpoint of the rural community," Benton wrote to the managing partner in Cincinnati in April 1915, adding, however: "I do not believe that Hunter is qualified to do the work."[16]

In 1920, Hunter approached Assistant Editor in Chief W. W. Livengood about the possibility of revising the *Essentials*.[17] Hunter had just moved to Galesburg, Illinois, to become a professor of science education at Knox College.[18] Perhaps he had planned to begin work on a revision before the new school year began, but it was not until the end of the summer that the ABC decided that an updated *Essentials* would be useful. Jarvis Fairchild and other partners were supportive,[19] but salesmen from the southern states were instrumental in bringing this about. One of them wrote:

> [The *Essentials*] has been successful, and has served its purpose, but has now to a great extent been succeed [*sic*] by the "Civic Biology".
>
> However, in this territory there has been a more or less well-defined tendency to differentiate the books between rural schools and urban schools, by pushing the "Essentials" for the former and the "Civic" for the latter. If, therefore, a revision of the "Essentials" be decided on, I would strongly urge increasing and emphasizing the rural content and applications of the book.[20]

Hunter did not actually begin work on the revised *Essentials* until late 1921. Livengood remarked to Fairchild: "Apparently he is either working on the revision or working himself up to the point of beginning to commence. He says he is in the middle of it."[21] But production quickly stopped when Livengood discovered that Hunter was commissioning artists to create illustrations for the textbook without going through the ABC's art department or getting approval for the outside expense. The exchange quickly grew heated, with Livengood informing Hunter: "If you want to take the entire responsibility of having the illustrations made without any supervision on our part, you will do so at your own expense."[22] Hunter angrily responded: "What I want is co-operation. If after twenty years of experience with me, . . . the American Book Company does not have more confidence in me than this, then I think it is either time for the Editorial Department to wake up or for me to take my wares elsewhere."[23]

The delays and conflicts that accompanied the creation of the *New Essentials* foreshadowed what was to come with the post-Scopes revision of the *Civic Biology*. Hunter already had long-standing relationships with Benton, Livengood, and many other editors and partners at the ABC, but those relationships were frequently complicated by whatever projects were going on at the time. The *New Essentials* was published in late 1923—years after it was first conceived. Hunter's work schedule and delays brought on by conflicts with his editors were likely to blame for the textbook not coming to market sooner. Had it come out sooner, perhaps it would not have contributed as greatly to the decline of the aging *Civic Biology*. Moreover, until the *New Essentials* was complete, neither Hunter nor his editors could have begun to discuss updating that 1914 book.

CREATING A NEW *CIVIC BIOLOGY*

On February 11, 1924, Hunter wrote to Benton about his plans for a revised *Civic Biology*. Hunter seemed to assume that this was a natural next step now that the *New Essentials* and its ancillary material were finished, but no one in the ABC offices knew what he was talking about.[24] This does not seem to have deterred Hunter, who wrote in June that he planned on spending the summer of 1924 revising the *Civic Biology* and wanted some copies of the book to cut and paste into the new manuscript. This spurred the ABC into action to try and decide whether a revised textbook would be useful and whether it even wanted to do another book with Hunter.[25] Benton solicited the opinions of the sales managers in the regional offices, trying to get a feel for whether there was a market for a new book. The replies were mixed. The *Civic Biology* had been a good book but was hopelessly out of date. Its sales were falling, and the *New Essentials* only accelerated its decline. If it could be released soon *and* it would not conflict with the *New Essentials*, then there would be a great demand for a new *Civic Biology*. Having seen the effect of long delays in producing the *New Essentials*, all agreed that time was of the essence.

The thing to do, the partners decided, would be to make the new *Civic Biology* a tenth-grade book. This would not only prevent it from competing with the *New Essentials*; it would also allow salesmen to present it as part of a series beginning with Hunter and Walter G. Whitman's *Civic Science in the Home and Community*, a ninth-grade general science book.[26] Just as the *New Essentials* had been expected to emphasize its rural character, this new textbook must be developed in such a way that it was explicitly marked as a book for city schools.

While the partners were debating this, Benton discussed with Hunter the vision the company had for the new book. He also spoke to Hunter about the antievolution concern. In September 1924, he told Hunter that a revised book "need not necessarily call for treatment of the theories of evolution, but should rather emphasize the relations to human health and conservation in society."[27] Benton did not record what Hunter's reaction to this was. On October 31, he told Hunter that the board of directors had formally authorized the revision of the *Civic Biology* and sent him a contract.[28]

Hunter wrote back that he would not be able to start the revision right away: "It is probable that I will be able to begin work on this sometime in the early spring." He was taking an unpaid semester's leave from Knox College to recover his health after coming "rather close to a nervous breakdown."[29] He planned to spend this time in La Jolla, California, and despite the delay felt he could finish the book by the late summer or fall of 1925. Benton replied that Hunter should look to his health, and he expressed hope that the book would come out in early 1926.[30]

Work proceeded slowly in the spring of 1925, in part because letters took a week to travel between New York and California rather than the two or three days between New York and Galesberg. Hunter began revising the content of the book in April, warning the editors: "I have two more months to work under pressure. . . . It is now or never if I am to do a good job."[31] Hunter's leave of absence from Knox College was set to expire on July 1, and he would begin, full of self-proclaimed vigor, to resume his teaching.

Throughout the 1924–25 school year, Benton and Livengood were beset with letters from their sales managers asking when a new biology textbook would be available. The decision in Tennessee to postpone a new adoption for a year had kept Hunter's book in classrooms there a little longer. Even though most people in the state did not need to buy a new copy and the ABC would not make much profit from those few sales, the postponement gave the company extra time to bring a new book to market. But visions of competing for adoptions in the summer of 1925 had already faded by May. Even if Hunter made good on his promise to finish his manuscript by the end of June, it would be difficult for the manuscript to be edited and the books to be printed and bound by the start of the 1925–26 school year. Sensing this urgency, Hunter offered to make a quick update of the *Civic Biology* that would require only a few new plates and could be used as a stopgap while the full revision continued.[32] Ginn's 1924 reprinting of Benjamin Gruenberg's *Elementary Biology* had performed precisely the same function, coming just a year before his new textbook was released.[33] However, Benton and the sales managers decided that it would be better for Hunter to devote his full energy to producing the new book, afraid that the distraction of even a minor revision would prevent him from ever finishing the new *Civic Biology*.[34]

Even though news of Scopes's indictment reached Hunter and Benton in May, it seems they both paid it little heed aside from the latter's initial caution not to speak on the matter. Much more pressing for Benton was news that the sales force had convinced Detroit to hold off on replacing its biology textbooks with the promise that a new *Civic Biology* was imminent. Hunter was determined to finish the book by the beginning of July so that he could travel through California and other western states before heading back to Galesburg. Benton despaired of Hunter's work ethic, telling another editor: "Mr. Hunter does not do very careful work when he is in a hurry, especially when he is hurrying to get away on a vacation tour which I understand he contemplates."[35] Toward the end of June, Benton wrote Hunter:

> I have another urgent call from the Chicago Office calling attention to the importance of getting the revision of Civic Biology on the market by January 1, 1926. There seems to be a feeling in that center of activity that if we do not come out early in 1926 with the revised book, we shall lose a large share of the civic biology business to some of your competitors.
>
> Please sacrifice everything except life and health themselves to this purpose.[36]

Despite giving Hunter this ultimatum, Benton had already written the chief of the ABC's manufacturing department: "There is no reason to expect that [the revised *Civic Biology*] will appear until rather late in 1926. . . . [I]t will not be issued until January 1, 1927."[37]

Hunter was not the only one whose summer vacation was affected by his progress on the manuscript. In early July, Livengood confided to Louis B. Lee (who ran the Chicago office): "I am quite sure . . . that he won't have his revision of Civic Biology in in time. . . . If he does, I shall be inclined to fall dead in my tracks, and since I am going to go away on my vacation, I have no desire to commit suicide, therefore I am not asking about the Hunter manuscript."[38] As it turns out, Livengood was wrong. Hunter had turned in his manuscript to the ABC's San Francisco office on June 25 as he left for his own vacation.[39]

Despite the news about Scopes during May and June, the topic of evolution and the pending trial had not been that important as Benton and Hunter discussed the revision. There was a lot to change from a book published in 1914. Not only had the curriculum of civic biology in lieu of distinct classes of botany and zoology become widely accepted in the intervening eleven years, but also new biological discoveries and new political circumstances meant that the revision would have to meet different student needs. Throughout May, Benton and Hunter discussed how the issue of alcohol consumption should be treated in the book. Prohibition had gone into effect in 1919; the need to teach young people about alcohol's harmful effects was not rendered moot, but the need to discuss public regulation of alcohol was certainly different. Benton noted that the 1914 *Civic Biology* spent nearly twenty-one pages across four chapters on the effects of alcohol, while the 1919 *Elementary Biology* by Benjamin Gruenberg treated the subject in only thirteen pages across two chapters.[40] Benton and Hunter looked to several other books for ideas of what was successful, though Gruenberg's books seem to have been the most influential.

The Macmillan-published *Biology and Human Welfare*, by Peabody and Hunt, was also important as it was the book that had replaced the *Civic Biology* in Tennessee, ostensibly because it did not discuss evolution. Reasons for that omission were given in the preface: "The field of biology is so large and some of its problems are so difficult that a wise selection of topics must be made if a text is to attain maximum usefulness. Consequently some topics that are often found in secondary school texts, but that are more suitable for discussion in advanced courses, have been omitted. Among these may be mentioned Mendel's laws, theories of evolution, and mitosis."[41]

The preface included a footnote that quoted from the College Entrance Board and the New York State Department of Education to the effect that theories of evolution were too advanced for high schools but that the evolution of animals within groups should be taught. On the day that Tennessee replaced his book with Gruenberg's *Biology and Human Life*, Hunter wrote to Benton: "Peabody and Hunt were lucky enough or crafty enough to take a quotation that made them safe in Tennessee. I think it will be well to modify the statement . . . in the Civic and make it clear that there is no intention to state that man descended from a monkey."[42] A week later, Benton pointed to *Biology and Human Life* as another example of an appropriate treatment of the antievolution issue:

> The Gruenberg treatment of the question in his latest book is very canny. I found a reference to the origin of species and to natural selection, but I failed to find the word "evolution" in the book. This of course is decidedly "pussy-footing" but it is not open to real criticism against the subject from the standpoint of schoolbooks which, as I get it, is aimed at the fact that even the discussion of such questions . . . tends to increase their lack of belief in and respect for spiritual things. . . . In these days of intense unrest and almost carping criticism on all subjects, I think we would do well to avoid overdoing this subject in a schoolbook.[43]

Hunter received this letter in San Francisco, where he had just dropped off the finished manuscript. He sent off a quick reply:

> The manuscript is complete, except in such details as the alcohol question and the illustrations. . . .
>
> You will note that I have changed the treatment of the evolution question to an extent, but have not "pussy-footed" as Gruenberg did. I talked with him on the matter the other day and he admitted that he had purposely left out the term evolution in the index. And of course Peabody and Hunt did a still more canny thing. . . . I have taken a middle ground. While I do not believe in teaching evolution from the discussion standpoint, I do not see how we can escape indirectly teaching it if one refers to classification and to the place of man in nature. Man, the mammal, will admit of no other interpretation.[44]

The original *Civic Biology* never explicitly stated that humans had evolved from nonhuman precursors. However, it explained that physiological similarities (not common ancestry) justify classifying them among the mammals. It also discussed the evolution of modern humanity from "men who were much lower in their mental organization than the present inhabitants."[45]

Hunter thought that it might be useful to explicitly emphasize that humans did not directly evolve from monkeys. In the revised manuscript he sent off, he had added the sentence: "It must not be considered that man evolved from a monkey."[46] The representation of human evolution as a claim that humans descended from monkeys was widespread, often becoming a sort of visual shorthand for evolution theory in the 1920s.[47]

Adding this sentence seems to have been the only gesture Hunter made toward addressing the evolution controversy. He seems to have thought that the whole flap over the teaching of evolution was a misunderstanding. He correctly argued that his book did not teach human descent from nonhumans and assumed that greater clarity should resolve the matter, but he was also insistent that the revision should not reduce discussion of evolution: "After the smoke of battle is over we will have more holdings than we had before; albeit we have lost Tennessee. . . . The new book will give no direct treatment of evolution as such, but the child who thinks cannot escape the conclusion that changes are going on in nature today, as in the past."[48]

After spending a few days in San Francisco to sort out some of the illustrations for the book, Hunter left for Portland, Oregon, while the Scopes trial was starting. From there he went to Glacier Park, Montana, for the first two weeks of August, with letters from Benton chasing him around the country.[49] While Hunter was traveling, the Scopes trial was remaking the evolution debate. William Jennings Bryan had risen to attack Hunter's book. John Scopes had been convicted, and the *Civic Biology* was cast as the textbook that taught the subject that rendered him guilty. Thanks to the trial, people understood the Tennessee law as not just banning a specific detail of human ancestry but as prohibiting anything to do with evolution. The Scopes trial also situated the idea of evolution, and by extension Hunter's *Civic Biology*, in open conflict with religion. By August, a popular understanding of the 1914 textbook was that it had caused the Scopes trial, taught evolution, attacked Christianity, and even indirectly led to Bryan's death. While this radical change in perception took place, the book's author, George William Hunter, was on his way to the Montana woods to rest his nerves.

It did not take long for other publishers to take advantage of the *Civic Biology*'s association with the trial. By mid-August, Louis Lee was reporting that sales agents from competitors were using quotations from the Scopes trial to discredit Hunter's book and promote their own.[50] Hunter wrote to Lee from Glacier Park to say that he did not see why the *Civic Biology* should suffer any more than other biology textbooks: "Moon, of course, treats evolution in one chapter openly. The new Smallwood, Reveley & Bailey book I haven't seen. The Atwood treats evolution openly. I do not see how our competing books can get away with anything on us." He also insisted that his book did not directly teach evolution, only indirectly making use of it: "To say that the civic is an evolutionary textbook from cover to cover is a plain LIE, invented for the purpose of cutting our sales."[51]

Hunter was missing the point. It did not matter whether other textbooks were actually teaching as much evolution. What mattered was how they were perceived. He may also have not understood how much the trial had changed the perception of the *Civic Biology*. But in the first few weeks after the trial no one in the ABC really knew how the antievolution matter would resolve itself. Having survived the arrival of Hunter's manuscript, W. W. Livengood went on vacation in July just before the Scopes trial. Other members of the editorial office went away in August. "Vacation absences and consequent congestion" slowed productivity in New York for much of the month.[52] Meanwhile, Lee in Chicago, Howe in Cincinnati, and the other regional office managers were canvassing their sales agents and were the first to discover the harm the trial had done to the *Civic Biology*'s reputation. Regardless of whether the book's content was actually more or less evolutionary than its competitors, its prospects—already bad in 1925—were now even worse.

Hunter's manuscript reached New York just after the trial had ended.[53] It was in remarkably bad shape. Poorly handwritten notes and overt haste made for a "careless" manuscript, one that failed to deal with one of the more serious concerns—the topic of alcohol consumption—at all.[54] Part of this was, as Benton suspected, that Hunter did not do good work when hurrying off to a vacation. But Benton had also told Hunter to hurry and to cut out most of the alcohol section, with the thought that the editorial office could restore some of it later if needed.[55] The editorial office, beset with pleas from their sales managers to hurry the manuscript, sent it out for review almost immediately, without even taking time to edit it first.

By August 3, the first outside review of the manuscript, from Annah P. Hazen, head of the Biology Department at Eastern District High School in Brooklyn, had been received. Hunter had recommended Hazen as a good person to review the book, calling her "very accurate and painstaking."[56] "The revision is a great improvement over the old book. There are many changes for the better," Hazen began. "The book is not just like all the other biology books which have gone before. It is somewhat original in its treatment and brings a freshness to the subject which seems likely to be a success." Hunter's style of writing and the organization of the book were particularly appealing. Yet Hazen also noted: "The word evolution is used on page 283 and the meaning of the term is explained on pages 311 and 312. Evolution is described without using the term on page 467. Darwin and his teaching of the theory of evolution is outlined on pages 498 and 499. Wallace and his contribution is explained on page 500."[57] So soon after the end of the trial, the presence of the word *evolution* was already being used to anticipate how readers would react. Perhaps this was due to the influence of Ginn's removal of the word from the index of Gruenberg's *Elementary Biology*.

Benton took his own vacation in early September, and his letter to Hunter on returning stated: "I will look into the present condition of the manuscript just as soon as I can get around to it."[58] This reply led Hunter to fret that not much had been done since he finished the manuscript in June.[59] Benton had warned Hunter that the ABC would save time and money by perfecting the manuscript as much as possible before setting it in type. Shortly before the New York office received Hunter's manuscript, Benton had guessed that typesetting might begin in September.[60] But the condition in which the manuscript arrived made that unlikely.

More progress had been made figuring out what the public perception of the evolution controversy would mean for textbook sales. Benton had one of the editorial assistants prepare a list of all the textbooks the ABC published that either made mention of or accepted the theory of evolution.[61] Sales agents, especially those connected with the Chicago and Cincinnati offices (which managed sales for the Midwest and the South), had solicited opinions and were reporting back to Benton and Livengood what the public reaction to the antievolution trial would mean for textbook adoptions in their districts. While Hunter urged Benton to hurry—"unless the book is in [the sales agents'] hands for the spring trade the holdings of the 'Civic' will be practically lost in many parts of the country"[62]—McHenry Rhoads, the superintendent of public instruction for Kentucky, was reading through the 1923 *New Essentials of Biology* and picking out objectionable passages:

> While you may be right in your statement that this is one of the least objectionable texts on biology, I am taking the liberty to call your attention to a statement that could have been omitted and made it still less objectionable. Near the bottom of page #311, under the title "Evolution of Man" the following words from the second sentence could easily have been omitted without endangering the truth,—"we find that at first he must have been little better than one of the lower animals". If that phrase had been omitted and the second and third sentences had been run together . . . it would have been much better.[63]

The phrase that Rhoads wanted omitted had also been used in the 1914 *Civic Biology*.[64] In addition, Rhoads stated: "Illustrations of the evolution of the horse is evidently somewhat fanciful."[65]

Just as Hunter was complaining to Louis Lee that nothing was being done by Benton to edit the new manuscript, Lee was writing to Benton that the Scopes trial had nearly finished off the ABC's presence in biology. Without having a new textbook very soon, and without extremely cautious treatment of the evolution issue, ABC salesmen would have no chance: "First; the furor that has been kicked up over the country concerning this book, growing out of the Tennessee case, all of our competitors are using against the book, and next spring they will double up on us and try to put our present book off the market entirely. If we therefore could have a new book by Hunter guarding very carefully this trouble stirred up in Tennessee, we could meet our competitors on an even ground at least, and probably a more advantageous ground, for the merry fight."[66]

A few days later, Lee received a review of Hunter's manuscript that he had solicited from Mary E. Robb, a biology teacher at Hyde Park High School in Chicago (where Hunter had taught a quarter century earlier). Rather than a general review, her report was a series of responses to queries posed by Lee. The responses were generally favorable but confirmed Lee's midwestern concern that the urban examples listed in the text too heavily represented the East Coast. While not objecting to the treatment of evolution, Robb noted that Hunter "speaks here as if evolution were an accomplished result instead of an unfolding process. As if it were the result of an outside force instead of being inherent in life itself."[67]

By the end of October, with hope of having the book ready by early 1926 fading rapidly, Benton proposed the unusual—and expensive—step of putting the manuscript into type as soon as Hunter incorporated the changes suggested by his reviewers and finishing the editing at the proof stage. "This means a modification of our usual procedure in the acceptance of manuscripts," Benton wrote to the managing directors.[68] It was a risky plan. The additional expense would save time and also, it was hoped, increase the chance of having the book ready for adoptions in the summer of 1926. This plan was approved, and Benton quickly apprised Hunter of the news, returning the manuscript for his revision on November 5, 1925. He cautioned Hunter:

> While we do not want to teach things that are not so, we shall probably insist that there are some things which may be so but which can well be omitted from a book of this grade. I see no reason why we should deliberately put into a book of this grade a lot of stuff about evolution and the origin of man in such a way as to be offensive to a very large market, which we, and I'm sure you, would like to reach. It may be an offense to your pride as a scientist to have to omit the airing of your views on these mooted questions, but I think I know you well enough to be sure that, if the omission of unnecessary and controversial matters will have the effect of increasing your income, you will be in full acquiescence!

Benton's tone reflected the urgency impressed on him by the sales agents, supplemented by the reports by outside reviewers that showed the unlikelihood of any kind of discussion of evolution being acceptable. Faced with Hunter's resistance and his erratic work ethic even when compliant, Benton closed with an ultimatum: "I do not mind saying that, if you do not use a good deal of judgment in editing out these things which cause trouble, we may have to do it ourselves, and we would rather you would do it."[69]

But Hunter did not receive Benton's suggestions with "full acquiescence" and replied immediately:

> To take out of this book all allusion to evolution would be like skimming the cream from the milk. It would be absolutely foolish and diametrically opposed

> to the whole scheme of revision I had in mind. For you to make the statement "I see no reason why we should deliberately put into a book of this grade a lot of stuff about evolution and the origin of man in such a way as to be offensive to a very large market," is in extremely bad taste and a rather foolish statement. If I had thought that you believed what you say in this paragraph, my confidence in you as editor-in-chief would be sadly shaken. . . .
>
> If you have any specific recommendations as to changes in this part of the book, I am perfectly willing to listen to them. If the agency force have cogent reasons for making specific changes I am willing to listen to them. But I am not willing to have you make a veiled threat that the editorial rooms will delete and re-edit this book for me when I know at the present time that you have no one in the editorial rooms who knows enough to do this with any degree of success.[70]

Benton tried to play down Hunter's accusations:

> Apparently, your health is quite restored. And I certainly congratulate you, for your letter shows evidence of very strong convictions and your well-known perseverance.
>
> I am quite as convinced as you are of the absurdity of the fundamentalist attack upon science, but I am also just as fully convinced that some of our scientific friends are inclined to lean over backward in their efforts to stir up mud unnecessarily. I think you belong to that class. . . .
>
> There is just one other thing that I want to say, and that is that a good deal of time may be saved by cutting out personalities and threats.[71]

To this last sentiment, Hunter replied: "Your last paragraph tells the truth, but *you* started it."[72]

While personal acrimony and frustration simmered, genuine revision was also being done. Hunter fixed the discussion of alcohol consumption and sent the first 124 pages of the manuscript back to Benton on November 20.[73] None of this material dealt with evolution.

Benton held the Ginn-published *Biology and Human Life* as the exemplar of how to deal with the issue, most importantly, removing mention of the word *evolution* (as well as *evolve* and similar variants). That this was the right approach was made clear by the book's success. It had been added to the Tennessee list of adopted books, and Ginn salesmen had been among the most vocal in using the Scopes trial to attack the ABC. In Florida, the Bible Crusaders, a Fundamentalist advocacy group, saw the replacement of Gruenberg's 1919 *Elementary Biology* (which made frequent use of the word *evolution*) with his 1925 *Biology and Human Life* (which did not) as "victory number one" in the fight against evolution.[74]

Hunter disagreed with Benton's plan to ape the Ginn strategy, observing: "[Gruenberg] was clever enough to make this change before the trouble came. If we make changes now, which indicate that we were wrong, it will weaken our position . . . and we will lose even more than if we maintained a dignified position."[75] Hunter, who had once offered to make a quick update of the *Civic Biology*, now accused his publisher of being too concerned with the short term. He argued that, even if they did not compromise and sales in the coming year were affected, the textbook would be vindicated when antievolutionism died down. The truth is, the editors were thinking about the long term but saw the issue differently. It seemed unlikely that antievolutionism would subside quickly with groups like the Bible Crusaders planning new pushes. In January, Benton received a news clipping from the Boston manager Jarvis Fairchild detailing the Bible Crusaders' efforts to remove the teaching of evolution from Florida schools. Fairchild expressed hope that Hunter's new book would not arouse the attention of this or any similar group.[76] Even if the antievolution movement did lose momentum, the length of textbook adoptions meant that its influence would be felt for several more years. The ABC had already lost Tennessee in 1925, and other adoptions were upcoming in the summer of 1926. Rather than taking the hard-line position and holding firm on evolution, the editors seemed resolved to follow the path others had taken.

Hunter sent in full manuscript in the beginning of December, protesting: "If you wish the book without evolution, then you will have to get someone else to write it."[77] The ABC board immediately approved it for publication,[78] and Hunter signed the contract for the book. The manuscript was now out of his hands. Even though the editing was not yet complete and some illustrations remained to be made, galley proofs were sent to the managing directors and, through them, to sales agents to solicit feedback.[79] At this point, there was hope that the book would be available by April 1, 1926, and could be offered for adoptions at the end of the school year.[80]

Communication with Hunter had by and large passed out of Benton's hands and into those of Assistant Editor Stiles A. Torrance. The change in editors heralded a promising start to the new year, and Hunter was delighted when Torrance sent him a limerick along with a letter discussing revision of the manuscript.[81] Hunter accepted some editorial suggestions from Torrance with no hesitation and disagreed with others in a more polite tone than he had used in his correspondence with Benton. He seems to have thought of Torrance as an ally, writing to him: "It is obvious that some of my changes are not to be allowed by the editor in chief. I confess to be much disturbed by this for it makes me feel that the book will not serve the purpose for which it has been made. There is such a thing as emasculating the text."[82]

That Hunter had earlier confided in other ABC employees, such as Louis Lee, seem to indicate his belief that, if he could convince *someone* at the ABC to take up his side, Benton would be prevented from "emasculating" his textbook. In fact, Lee forwarded Hunter's missives straight to the editorial office. "I am enclosing a copy of a little statement from our friend the Honorable George W. Hunter, Knox College, Galesberg, Illinois," he jested. "I suppose he wrote me because he was in here the other day, on his return from California."[83] Livengood also seemed to mock Hunter, once even telling him: "I did not reply . . . because I was too busy to engage in petty bickering."[84] Later, when W. T. H. Howe had become ABC president and Livengood editor in chief, Hunter tried to write to Howe directly about a matter concerning which he felt Livengood had not given his opinion a fair hearing. Howe promptly gave the letter to Livengood to respond to.[85] Stiles Torrance was more congenial (even as Hunter repeatedly misspelled his name). But he was no less focused on the ABC's aim of producing a marketable textbook as quickly as possible.

In reality, Hunter's greatest champion within the ABC was probably George Benton. As the book neared completion, pressure grew from Lee and Howe in Chicago and Cincinnati to hurry and to further remove evolution from the text. Benton resisted these efforts as much as possible.

This became harder in January 1926, when Howe received a review of the new manuscript by Jesse M. Shaver. Shaver was the professor of biology at Peabody College in Nashville, Tennessee. Howe had even had Shaver paid "an extra fifty dollars" for the review "in order to get his active influence in the Peabody school and the city of Nashville."[86] Howe recognized that Peabody was quickly becoming the center of teacher training and education research in the South and thought it wise to develop good relationships with its faculty.

The first part of Shaver's report made its way from the Cincinnati office into Benton's hands on January 19, 1926.[87] It was thorough and pragmatic. The accompanying letter was to the point:

> From the standpoint of securing adoptions of this book in the southern field, it would be better to omit altogether, paragraph (Charles Darwin and Natural Selection) 6, page 206, galleys, and topic, Charles Darwin, pages 221 and 222, galleys.

> I believe it also undesirable to include in the new biology of Hunter, the diagram of "The geological history of the horse," of the old, *A Civic Biology*; diagram, "The evolutionary tree," page 194, old text; picture, "The four-toed ancestor of the present horse," page 260, old text; and the picture of Charles Darwin, page 404, old text.[88]

In lieu of the picture of Darwin, Shaver suggested one of Gregor Mendel. Attached to the letter was a five-page report listing page, paragraph, and line of suggested omissions and amendments. Of the forty-five specific suggestions made by Shaver, sixteen concerned evolution or Darwin. These repeatedly anticipated what "anti-evolution people" would do: "Some sentences here will cause anti-evolution agitation." "To make more acceptable to anti-evolution people, substitute. . . ." "Certain parts might cause anti-evolution controversies." Other comments addressed factual errors regarding animal physiology or other uncontroversial topics. In Shaver's second report, sent in February, all twenty-seven of his suggestions concerned evolution (some of these overlapped with suggestions made in the first report).[89] Shaver suggested excluding from the bibliography any reference book containing the word *evolution* in its title. He also suggested revised versions of entire paragraphs.

Personally, Shaver favored teaching evolution and did so at Peabody College. Nonetheless, he was able to instruct the ABC as to the adoption environment in Tennessee and the South after the Scopes trial. He occupied a position of authority in science education in the South. His endorsement of the book would mean a tremendous boost in sales. Howe's investment paid off when Shaver authored a review of the *New Civic Biology* in the *Peabody Journal of Education* in 1929. "There is no discussion of evolution," the review states bluntly. "This probably makes the book better suited to the conditions of our Southern high schools."[90]

When Shaver's first report on the Hunter manuscript reached New York, Benton met with both Torrance and Livengood to speak with them "especially about evolution and Darwin."[91] It was at this stage that Benton proved himself to be Hunter's staunchest defender in the ABC. He wrote to Hunter of "the problems with which you and Mr. Torrance and the rest of us are wrestling, that we are attempting an almost impossible thing in the revision of this book": "We are attempting it because we believe that it is desirable to get the book done in less time than is usually necessary to produce first class work." Benton was sympathetic to Hunter's concerns, but he had to weigh them against the needs of the ABC's sales force and his understanding of how textbook adopters were responding to the antievolution movement. "No one here is personally out of sympathy with your liberal views," he wrote Hunter, "but we are in the market to sell books."[92]

Howe continued to press the editorial office to incorporate all Shaver's recommendations, even suggesting that a quick edit of the original *Civic Biology* be rushed out with Shaver's changes. Benton received this suggestion coolly: "Just at present I am inclined to think that we ought not to make any changes in the original book, but we certainly can take advantage of these criticisms to see that the same things do not appear in the revision."[93] He also told Torrance and Livengood: "I am inclined to go slow with changes in this book. We cannot afford to drop Evolution altogether."[94]

Even so, Benton was receptive to Shaver's suggestion that the word *evolution* be removed. Torrance informed Hunter of this in early February 1926, and Hunter lamented that Benton "intends to soft-pedal all the evolution references." Convinced of this, Hunter declared: "I cannot fight against the entire office force and am inclined to turn the correction of proof over to the editorial rooms and forget that I am attempting to edit it."[95]

From Cincinnati, Howe continued to push for a removal of evolution from the book. "During this furor I do not believe we should admit we are wrong and trying to put something over by revising the present book," he wrote. But he held that the ABC could issue the new book without explicitly claiming it as a response to the trial.[96] Benton continued to have reservations about doing this, but at the same time more of Shaver's recommendations were incorporated.

In March, Hunter visited Chicago and met with Louis Lee and ABC president Louis Dillman. Afterward, he wrote to Benton. The Chicago manager had suggested that there were words that could be used as alternatives to *evolution*. Hunter agreed that other words could be used but noted: "I did not intend to have him think that I agreed entirely with him regarding the deletion of the term." While in Chicago, Hunter had also met with Mary Robb, the biology teacher who had provided one of the original reviews of the manuscript. According to Hunter: "She said she would prefer to see more evolution in the book. She did not, however, make this comment for the editorial rooms because she misunderstood Mr. Dillman to say the book company wanted to soft-pedal the evolutionary material."[97] Hunter closed the letter by stating that he would come to New York in early April and would speak to Benton personally about the matter.

Around the same time that Hunter visited Chicago, the managing directors settled on the title of the revised book, the *New Civic Biology*. But even this seemingly simple decision was contentious. Some—especially southern sales agents—had felt that the word *civic* had been too tarnished by association with the trial.[98] Hunter strenuously argued in favor of retaining it.[99] On this issue, Benton stood with Hunter against the opinions of his sales force: "I have read your letter of the 15th with varying emotions. What you say with reference to the title of the new book will, of course, be given careful consideration. . . . Perhaps your suggestion of a moderately changed Civic Biology with emphasis on evolution is what some people want and this suggestion also will have consideration."[100] Arguing that the book was intended to follow on Hunter and Whitman's *Civic Science in the Home and Community*, Benton convinced the other directors that the word *civic* should be retained.[101]

Hunter's emotions also varied—between anger and despair. "I have never felt so depressed and disgusted with a revision as with this one," he wrote Torrance in March. "I thought I had material for a mighty good book and it was before you people *spoiled* it."[102] By April, Hunter had just about given up. Proofs of the book were sent to him for final corrections, and he ignored them.[103] In early May, with publication "within a few days," Torrance wrote Hunter to express surprise that he had not received any corrections. "If you have any important alterations to suggest," he wrote, "please let us have them immediately."[104] Hunter replied a few days later, as the book was already going to press: "I have not turned in corrections before because I was so unutterably disgusted with the way the editorial rooms has handled the subject of evolution. I am enclosing, however, a number of changes none of which are vital but which should all be included in the second printing."[105]

As copies of the *New Civic Biology* rolled off the presses of the ABC's printing plant in New Jersey, Hunter had all but abandoned the book in revulsion. Benton seemed worn down from the months of trying to find a balance between his author and his sales force. The only parties who seemed to be happy about the new book were the salesmen. Howe, in particular, noted with glee that his salesmen had just received a review of the *New Civic* from a high school instructor in Alabama, which was just about to adopt new books. Even though it came too late to help revise the book, "as a matter of agency strategy" it would help in the upcoming battle for adoptions that summer.[106]

The *New Civic Biology* did away with any mention of the word *evolution*, just as Gruenberg's *Biology and Human Welfare* had done. Nearly all Jessie Shaver's recommendations were incorporated. Conciliatory statements regarding religion—for example, "Man is the only creature that has moral and religious instincts"—were added.[107] The images of the classification tree of animals and the fossils of prehistoric horses were gone.

Several reviewers had offered advice on how to negotiate the discussion of evolution in light of the Scopes trial. Even though most of those reviewers supported evolution, their suggestions presumed that the book's audience would not. Almost no one involved with the production of the *New Civic Biology* wanted to take evolution out of it. But somehow everyone assumed that doing so was necessary in order to compete in the South or anywhere else the antievolution movement might gain strength. The vilification of the *Civic Biology* during the trial helped create that understanding. So too did the successes of other textbooks before and after the Scopes trial.

The development of the *New Civic Biology* shows just how much a mistake it can be to attribute all the content of a published text to its author. The one person who never wavered in his commitment to keeping evolution in the book was its author, George W. Hunter. His insistence was not just the result of a "desire to preach evolution," as Benton once remarked.[108] He sincerely believed that, by pandering to a southern market that was already a lost cause, the ABC risked alienating the northern audiences who wanted evolution taught. Throughout the months of discussion and revision, he reminded Benton and others that he had talked with teachers across the country and that all of them wanted evolution to remain in the biology textbooks. But Benton kept in mind something that Hunter overlooked—that *teachers* did not make schoolbook adoptions.

Hunter was the author of the *New Civic Biology*, but by the time it was published he could not bear to look at it. Even though he lost many of the battles over revisions to the book, his efforts shaped part of the revision process. To the extent that the book was written—or rewritten—in response to the Scopes trial or the broader antievolution movement, its authors were people like McHenry Rhoads and Jessie Shaver, W. T. H. Howe and the sales agents, who suggested other revisions. To a lesser extent, its authors were also editors like Benton, Torrance, and Livengood.

It would not be correct to say that the Scopes trial itself compelled the rewriting of the book. The trial had no legal bearing on the textbooks used in Tennessee or anywhere else. But it did create a different *reading* of the *Civic Biology*. During the first few months after the trial, ABC employees learned how to reread the book in light of the new, posttrial environment. Then they began to rewrite it, using the reading lessons provided by members of their audience and the examples set by other textbooks. In part, the ABC's managers, especially Howe and Lee, learned how to create a book that could be interpreted as acceptable by an antievolutionist. Their salesmen were then charged with the task of selling the *New Civic Biology*. This meant selling not only the bound and printed object but also a way of reading the textbook as nonevolutionary. They had help: other textbooks that had already been adopted paved the way for acceptance of Hunter's new book. Jessie M. Shaver's review explicitly reassured education leaders in the South that the book did not contain evolution. So did the Alabama review that arrived just as the book was published. The *New Civic Biology* was adopted in places that were formally opposed to teaching evolution, demonstrating that the ABC was successful in teaching adoption commissions how to read the textbook in the way they wanted. The textbook did not sell itself; the ABC's agents, the favorable reviews of men like Shaver, and even Hunter's ongoing reputation as an author did.

Hunter's reading of his original *Civic Biology*—that it could stand some clarification but did not really teach the prohibited subject that humans evolved from other primates—was a valid interpretation of the text, but after the trial it was an irrelevant. His disagreements with his editors were partly due to his mistaken belief that what teachers wanted mattered. But, perhaps even more importantly, Hunter had largely missed the Scopes trial. He had been busy traveling, resting his nerves and away on vacation. He did not participate in the spectacle that had had such an impact on science and religion in America. It took longer for him to understand the changes the trial had wrought and what they meant for biology education. In effect, what he had missed was the reframing of the antievolution movement around the view of science and religion conflict that the trial itself popularized. Benton and the other editors and sales agents were more sensitive to this transformation and realized that it would somehow change the way in which textbooks could be read.

In a post-Scopes environment, antievolutionism had come to mean the prohibition of the teaching of *any* kind of evolution—not just human. The *New Civic Biology* was successfully presented as compliant with this new standard. Benton and his colleagues had also learned to include passages that seemed to show science and religion compatibility. The old *Civic Biology* arguably had not been in violation of the Tennessee law, but no one would have dared make that case after the Scopes trial. Ironically, nearly everyone accepted that the *New Civic Biology* was compliant. Perhaps some readers assumed that there was no way the ABC would release a new book by Hunter unless it was substantially changed and, thus, were inclined to see the book as nonevolutionary. But, reading the *New Civic Biology* closely, it is hard to see how its content was all that different from its precursor's.

Hunter maintained that the original *Civic Biology* did not teach evolution directly, making use of it only indirectly to teach about plants and animals

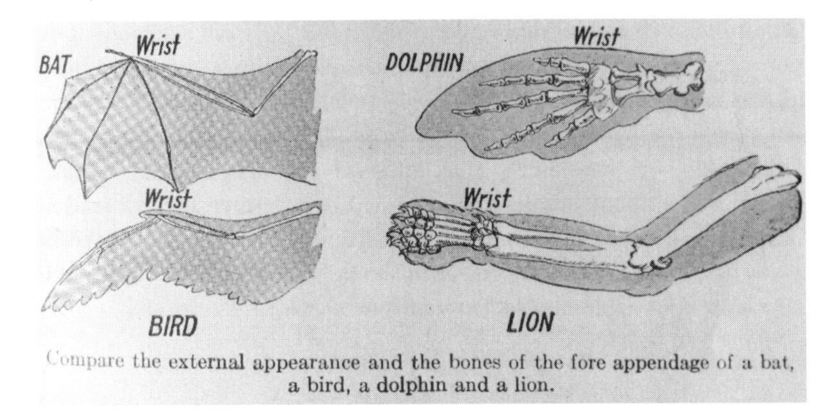

FIGURE 5 Even though human beings are included in the discussion of homologous structures in the text of the *New Civic Biology*, they are left out of the corresponding image.

and their relationships to human society. The *New Civic Biology* appears to do the same thing. For example, the new book pointed out homologous structures such as "the wing of a bat, the wing of a bird, the fore flipper of a dolphin, the fore leg of a lion, and the arm of a man" but characterized them as having merely "likeness *in function* regardless of origin."[109] As the text is phrased, it implies without explicitly stating that homology is evidence of common origin. Although it includes "the arm of a man," the illustration showed only "the wing of a bat, the wing of a bird, the fore flipper of a dolphin, [and] the fore leg of a lion" (see fig. 5). Unlike the 1914 textbook, the *New Civic Biology* actually does teach that humans share ancestors with nonhumans. Its discussion of life history removed the illustration of various horse fossils and the strata where they were located[110] but still teaches that "the earliest forms of life upon the earth were very simple and that gradually more and more complex forms appeared, as the rocks formed latest in time show the most highly developed forms of plant and animal life."[111] Despite the inclusion of the sentence "Man is the only creature that has moral and religious instincts" and the removal of the phrase "we find that at first he must have been little better than one of the lower animals,"[112] the discussion of the development of humanity from more primitive forms to that found in the modern day is also retained. In some passages, such as the biographical paragraphs on Charles Darwin, the text is almost entirely unaltered, except that the word *evolution* is replaced with the phrase *heredity and development*.[113]

For all Hunter's lamentation, the *New Civic Biology did* teach about evolution. How was it able to gain acceptance among those very same people who had so opposed the first *Civic Biology* (which arguably taught less about human evolution)? In the post-Scopes world, it was still true that textbook salesmanship was paramount. More attention was paid to content than in the past, but textbook adopters were taught by salesmen how to read and interpret that content. It was not just the ABC; other publishers took similar approaches. Removing the word *evolution* was a major step toward making their books acceptable in a society that saw the antievolution issue strictly on science-religion terms.

CHAPTER SEVEN

Biology Textbooks in an Era of Science and Religion

As the case of the *New Civic Biology* shows, the making of a textbook is often a complicated process. Even though one man's name—that of George W. Hunter—appears on the book's spine, many other people were involved in its creation. Readers of a book would likely assume that the object before them had been subjected to editing, and, while they might not know that George Benton, W. W. Livengood, and Stiles Torrance were the specific men involved, they would not be surprised to know that editors played a role in creating the text. The name of the publisher on the spine implies as much—that the American Book Company (ABC), which had a reputation for producing books of reliable quality, had vetted the words of George Hunter and adjusted them accordingly prior to publication. To look no further than the cover of the book already gives accounts of two sources of authority by which the *New Civic Biology* could be seen as reliable. Those two sources also indicate the collaborative nature of the process that produced the book. A book-consuming audience in the present-day United States would expect to find information about both author and publisher on any printed book.

But the school textbook contains other imprimaturs besides those that typical trade books contain. One of these is the mark of official approval by the state or local school district that has adopted the book. In cases where textbooks are loaned to students, these are labels indicating ownership—that a book is the "property of" a school. The school or state's willingness to purchase a book for students communicates to its readers approval of the book and its content. Even where a textbook is adopted but not purchased, there are indications that the book has been authorized. In 1920s Tennessee, state-adopted schoolbooks had a label affixed to the inside cover. Primarily,

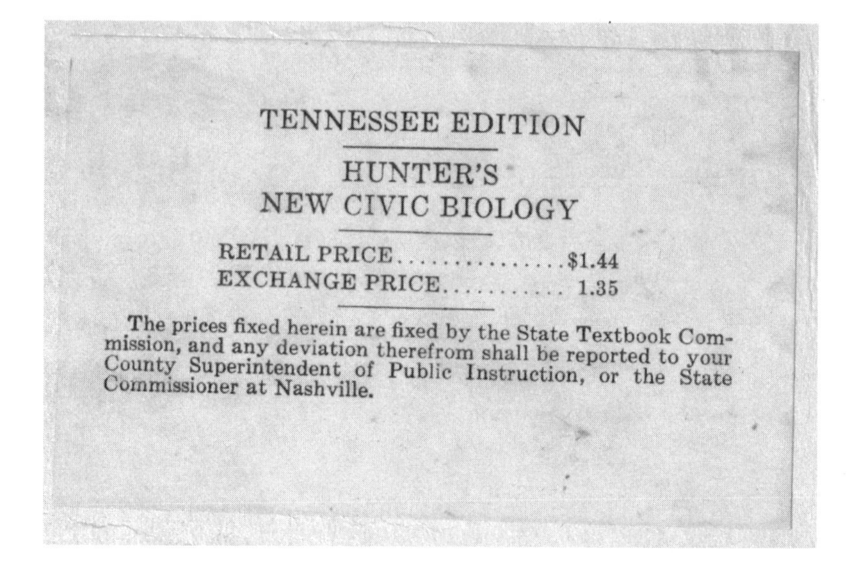

TENNESSEE EDITION

HUNTER'S
NEW CIVIC BIOLOGY

RETAIL PRICE................$1.44
EXCHANGE PRICE........... 1.35

The prices fixed herein are fixed by the State Textbook Commission, and any deviation therefrom shall be reported to your County Superintendent of Public Instruction, or the State Commissioner at Nashville.

FIGURE 6 Sticker inside copies of Hunter's *New Civic Biology* sold in Tennessee.

this was to control prices. The state label indicated the fixed prices of books to prevent distributors from overcharging (see fig. 6). Long before school boards were using stickers inside biology textbooks warning of evolutionary content,[1] the practice of physically attaching a label to textbooks was standard.

The physical traces of authorship and authorization give some indication of how a textbook is likely to be read and consumed by its audiences, but they give an incomplete view of the parties involved in creating the text. The state label indicates only that the book has been accepted by the state; it does not indicate the extent to which state standards, outside groups of readers, and the input of state adoption boards influence the writing of a book. Several parties who were neither the book's author nor its editors had invisible hands in the writing of the *New Civic Biology*. The high school teachers in New York and Chicago who commented on the manuscript signaled things that should be revised and made other suggestions. The Peabody College of Education biology professor Jessie M. Shaver and Kentucky superintendent of public instruction McHenry Rhoads did much to rewrite the book. Arguably, the overall book—not just its treatment of evolution—more closely resembled Rhoads's view of biology pedagogy than Hunter's.

The most significant difference in the *New Civic Biology* was that it avoided the word *evolution*. It was not the only textbook to do so, yet nearly all those that did still taught evolutionary concepts in one form or another. This does not seem to have been the result of any collusion among the rival publishers or among the authors themselves. Nonetheless, this technique came to be seen as a legitimate and successful response to the antievolutionary culture that prevailed after the Scopes trial. The way this came about illustrates the extent to which audiences can create the texts they consume. These biology textbooks were produced along with certain ways of reading them that allowed them to be a response to the Scopes trial. Even in some cases where the textbook was published shortly *before* the trial, the way in which these books were able to be sold and read evolved in such a way that they too could be seen as responses to the trial.

THE ORIGINS OF THE ANTI-"*EVOLUTION*" MOVEMENT

A month before the Scopes trial began, an editorial appeared in the *Peabody Journal of Education*. The Nashville-based school of education that W. T. H. Howe called "the Columbia of the South" had become even more influential with the simultaneous rise of public schooling in southern states and the founding of its own in-house publication in 1923.[2] The journal helped collect and disseminate information about issues in education of particular importance to southern educators. This editorial, "The Teacher and the Truth," lambasted the antievolution law that had just been enacted just two miles from Peabody's campus: "George Peabody College of Teachers and Vanderbilt University are not directly affected by the legislation, because they are not state-supported institutions; but both are deeply concerned, nevertheless, for their graduates who teach within Tennessee will have to contend with this mischievous and bigoted handicap to their highest service towards public intelligence."[3]

The editorial not only condemned the antievolution law; it presented an interpretation of the law that set evolution against specific understandings of particular Bible passages: "The edicts are directed against the teachers of biology, in whose classes the question of the origin and development of man's body naturally arises. 'So God created man in his own image' (Gen. 1: 27), is the key argument, the slogan of the opponents of evolutionary theory. And yet it is positively revolting to accept the insistent claims of literalists as to a bodily resemblance between man and God, and think of a Deity with a stomach, a liver, an appendix—out of reverence we go no further." As the editors of the *Peabody Journal* posed it, opposition to evolution was rooted in the absurd, "insistent claims of literalists." These literalists were defending, not the Bible, but "their private interpretations of it, in which they have so much confidence that they would foist them upon every one else."[4] Like other evolution supporters, the editors equated literalism with an interpretation of the Genesis account of creation. Invoking Andrew Dickson White, they claimed that history will eventually validate science, as it always has in its slow unfolding warfare with theology.

This editorial is typical of much of the anti-antievolutionary rhetoric before the Scopes trial. Were it not for the *Peabody Journal*'s particular audience of schoolteachers and other education experts (especially in the South), the editorial might be unremarkable among reactions to the antievolution law. However, it also suggests a likely scenario if the law were to be enforced: "Teachers would evade the law by teaching guardedly the facts, but never using the forbidden word, 'evolution,' or alluding to a monkey (in which case very possibly not one critic in a thousand could identify the forbidden hypothesis)."[5]

The idea that one could circumvent antievolutionary sentiment by simply avoiding the word *evolution* goes back to at least 1923, when Ginn and Company editor in chief Charles Thurber told William Jennings Bryan about his company's experience with antievolutionism:

> In one instance where one of our books had been barred from a certain institution we were informed that if we would merely cut out the word "evolution" the book would be reinstated. The statements of the book would remain just as they had been before except that for evolution we should substitute some other term which might mean as nearly as possible the same thing but which would not be known to the public and therefore was not objectionable.
>
> Naturally we want to sell our books, but we do not like to resort to such a trick or subterfuge as this to get business, nor do we feel that we should entirely suppress every reference to a philosophical and scientific concept which, whether we approve it or disapprove of it, has affected the life and thought of two generations.

Thurber noted: "It does seem to be true that to a great many people the mere use of the word 'evolution' is enough to stamp the one who uses the word as an atheist."[6]

In his reply, Bryan did not wholeheartedly endorse the removal of the word *evolution* as either necessary or sufficient to allay the concerns of antievolutionists. He did note: "The publishing house that recognizes the revolt against evolution will find it much easier to reach the authorities, especially in the South where a much larger percentage of the people are Christians." He also confirmed Thurber's understanding that (for Bryan at least) teaching evolution as a hypothesis was acceptable, "that it is the teaching of it *as a fact* that is objected to." He offered Thurber advice on how to create books that could be marketed in this controversial environment: "First, by eliminating objectionable phraseology from the text books. It would take a great deal in the way of elimination and addition to make it clear that evolution is presented only as an hypothesis, unproven, but an hypothesis that educated people should understand. If such a book is used as a textbook by teachers who do not themselves believe in evolution, the facts upon which evolutionists base their belief would be harmless, because it is not the facts that do harm but the conclusions based upon the facts—forced conclusions unsupported by fact."[7] Bryan did not recommend the simple subterfuge of removing the word *evolution*, but he noted that explaining evolution as a hypothesis and not as a fact would be very difficult.

Ginn took no immediate action, even when it had the opportunity to do so. In 1924, it released an updated edition of Gruenberg's *Elementary Biology*. The new edition corrected a few errors and updated some necessary details. To save money, care was taken to replace excised text with text of the same length and format. That way, subsequent pages in a chapter could be printed using the plates from the 1919 edition, and there would be no missing section or page numbers. The time and expense of creating new electrotype plates was a frequent concern for publishers.

But none of the changes in the 1924 revised *Elementary Biology* altered the book's discussion of evolution. Even when a new plate was already being made for a page discussing evolution—when it would have been effectively free to make a change—there was no attempt to downplay or alter the discussion.[8] It would have been difficult to remove evolution from the *Elementary Biology*; to do so would have required changing an entire unit of the textbook. By the time Thurber wrote to Bryan, Gruenberg's new textbook was already being planned.

A COMMUNITY OF REVISIONS

It was not until Gruenberg's *Biology and Human Life* was being completed that Ginn acted on Bryan's advice. The book was published just as Tennessee was passing its antievolution law. The word *evolution* does not appear at all in this book, yet one need turn only a few pages to find pictures of ancient plant and animal fossils with captions describing them as millions of years old. In the last unit of the book, along with chapters devoted to eugenics Gruenberg references the ongoing evolution of humanity, writing: "By saving lives, or postponing deaths, civilization interferes with natural selection."[9] There is less discussion of species change, heredity, or variation than the *Elementary Biology* contained; most of the discussion that is included is focused on applications of heredity to improve the human species or the plants and animals it uses. Unlike the way in which the *New Civic Biology* was finally revised, Thurber and Gruenberg seem to have worked together from the beginning in crafting a book that avoided "objectionable phraseology." Though not as explicitly evolutionary, the *Biology and Human Life* nonetheless presents as facts the antiquity of life on earth. It was added to the list of adopted books in Tennessee in June.

The textbook Tennessee adopted just before the trial, the Macmillan-published *Biology and Human Welfare* by Peabody and Hunt, mentioned the word *evolution* only in its preface, stating that the subject was not taught in the book. Walter White, the county school superintendent who testified at the trial that Scopes taught evolution, said of this new textbook: "It conforms with the law. You won't find any evolutionary tree in this book."[10]

Despite the disclaimer in the preface and White's reassurances, only a few pages into the book there is a laudatory description of Charles Darwin. A caption underneath his portrait called him "a painstaking investigator of problems bearing on the theory of the origin of living forms." The main text of the book states that it is "Darwin to whom the world owes a great part of its modern progress in biology." He is also credited as "publishing one of the epoch-making books of all time, *On the Origin of Species*."[11]

Gruenberg and Ginn had the benefit of making a new book in *Biology and Human Life*. Gruenberg intended this book to serve, not as a revision of the *Elementary Biology*, but as a tenth-grade biology as part of a two-market approach to the topic. (Several years later, it became clear that Ginn's sales offices were not marketing the books this way, and the *Elementary Biology* was virtually abandoned.) By starting from scratch, it was easier to do more than merely avoid mention of the word *evolution*. Some other books published in the year or so before the Scopes trial had already taken some steps to avoid the objections of antievolutionists, but the trial changed the criteria by which such books were being judged. Publishers could either wait until enough time had passed to warrant a full revision of a biology textbook, in which case they could, like Ginn, shift the overall focus away from evolution, or they could remove the obvious: the word *evolution*, discussion of Darwin, and depictions of evolutionary trees.

Macmillan's book had just been adopted in Tennessee, and the publisher made no immediate effort to alter the *Biology and Human Welfare* after the

Scopes trial. The ABC had considered making some quick changes to the *Civic Biology* but ultimately decided to have George Hunter focus on the manuscript of the *New Civic Biology*. Other publishers were in a more difficult situation. They could not abandon textbooks so recently published. Too much time and money had been spent in developing and promoting them to abandon them before their full lifetime had been spent. To have an author make a revision, even to make substantial changes in the editorial rooms, would also require a considerable investment. With new sales constantly looming, many publishers opted for the less thorough, less expensive, and less laborious revision.

Edward Southworth's Iroquois Publishing Company exemplified this approach with its biology textbook. The former Ginn salesman and textbook coauthor published his first biology textbook in 1924, Arthur Clement's *Living Things: An Elementary Biology*. The next year, Southworth was in Tennessee to pitch *Living Things* to the state textbook commission, and again he made a focused pitch. But it was not adopted in June 1925.

Perhaps anticipating the antievolution criticism, the 1924 edition of *Living Things* did not include much mention of evolution at all (though it included eugenics and sections on "the struggle for existence," "natural selection," and the "survival of the fittest"). The only place where the word *evolution* occurred was in the final chapter, which comprised a series of biographical sketches of famous biologists and included several paragraphs about Darwin. The discussion of evolution (shown in fig. 7*a*) did not try to insist on the theory's truth but rather took an almost apologetic tone: "Contrary to popular opinion, Darwin did not originate the idea of evolution. He only applied a very old doctrine and endeavored to prove that natural selection was the means by which evolution is effected."[12] A question at the end of the chapter and an entry in the index were the only other places where the word *evolution* occurred.

Southworth failed to win the book's adoption in Tennessee in June, but after the Scopes trial Iroquois made a quick and economical revision. Out of nearly five hundred pages, the revised *Living Things* required only five new plates. One was the copyright page, changed to emphasize to potential markets that the book was new. The other four changes consisted of the two pages on which the biographical summary of Darwin was found (see fig. 7*b*), the page of questions at the end of that chapter (one of which had included the word *evolution*), and the page of the index that had contained the entry for evolution (see fig. 8).

Considerable effort was made to ensure that the new paragraph on Darwin ended *exactly* two lines into page 456 so that plates of the subsequent

Darwin.—Charles Darwin was born at Shrewsbury, England, in the year 1809. As a boy he was interested in making collections of birds' eggs, shells of mollusks and other objects. He developed a strong interest in natural history, and after his graduation from Christ's College, Cambridge, in 1831, he received a position as volunteer naturalist on the Beagle, a ship sent by the English government on a trip around the world. During this voyage, which lasted five years, he made many observations on the forms and habits of plants and animals which form the nucleus of his most notable book, *"The Origin of Species by Means of Natural Selection."* This treatise shows accurate knowledge of facts of natural history and wonderful power of generalization. It has passed through many editions, both at home and in other countries, and owing to its advocacy of the theory of evolution has been the cause of much controversy. Contrary to popular opinion, Darwin did not originate the idea of evolution. He only applied a very old doctrine and endeavored to prove that natural selection was the means by which evolution is effected. According to Darwin, the most important factors involved in the origin of species are variation among individuals of a species, struggle for

455

456 Living Things

existence, survival of the fittest and inheritance of favorable characteristics.

Pasteur.—Louis Pasteur was born at Dole, France, in 1822. In his student days he made a special study of chemistry and

FIGURE 7 Pages 455 and 456 of the 1924 and 1925 editions of Clement's *Living Things* (figs. 7*a* and 7*b*, respectively). In the post-Scopes version, care was taken to ensure that the replaced text about Charles Darwin took up the same amount of space so that plates for the rest of the chapter would not need to be made.

pages of the chapter would not have to be remade. References to the *Origin of Species* and the word *evolution* were cut. Something needed to be inserted in their stead. That something turned out to be a lengthy list of books that Darwin wrote, not including the *Origin of Species*, the *Descent of Man*, and the *Expressions of Emotions in Man and Animals*. The paragraph is almost entirely filler, using drawn-out phrasing to say very little about Darwin's contributions to science.

... of living things.

Darwin.—Charles Darwin was born at Shrewsbury, England, in the year 1809. As a boy he was interested in making collections of birds' eggs, shells of mollusks and other objects. He developed a strong interest in natural history, and after his graduation from Christ's College, Cambridge, in 1831, he received a position as volunteer naturalist on the Beagle, a ship sent by the English government on a trip around the world. During this voyage, which lasted five years, he made many observations on the forms and habits of plants and animals. His writings show accurate knowledge of the facts of natural history and wonderful power of generalization. He wrote and published several books among which are: *The Voyage of a Naturalist; A Journal of Researches into the Geology and Natural History of the Various Countries Visited by H. M. S. Beagle; The Structure and Distribution of Coral Reefs; The Different Forms of Flowers in Plants of the Same Species;* and an interesting essay on *The Formation of Vegetable Mold through the Action of Worms.*

Darwin's whole life so far as his strength allowed was given to the interests of science and his work received many recognitions. He was made a fellow of the most important scientific

455

456 Living Things

societies of Europe and was honored with degrees from several leading universities. He died April 19, 1882.

Pasteur.—Louis Pasteur was born at Dole, France, in 1822. In his student days he made a special study of chemistry and

FIGURE 7 (*continued*)

The 1925 *Living Things* was fixated on removing the word *evolution*. Other evolutionary content (such as there was) was retained. But the decision to simply remove the word *evolution* was made, probably not because Southworth knew that this was what antievolutionists wanted, but because any more comprehensive changes would have taken too much time and money. Regardless, Southworth's revision was a success. In October, the Tennessee state textbook commission added *Living Things* to the state list, joining Gruenberg's *Biology and Human Life* and Peabody and Hunt's *Biology and Human Welfare*.[13]

Index 477

Embryology, 9.
Emergencies, 279-288.
Emulsion, 219.
Enamel, 222.
Encysted, 135.
Endosperm, 352.
Enemies, of insects, 59; of crayfish, 69; frogs, 90; birds, 99; plant life, 402.
Energy, 23-27, 33, 73, 151, 185, 235.
English sparrow, 97, 98, 99.
Environment, 40, 453.
Enzyme, 84, 135, 183, 211, 214, 215, 218.
Epicotyl, 352.
Epidemic, 172, 268, 411, 414, 420.
Epidermis, 237, 298, 324.
Epiglottis, 158.
Epithelial cells, 236, 237.
Epithelial tissue, 89.
Erect stems, 319.
Erosion, 361, 362, 400.
Esophagus, 74, 83, 211, 213.
Essential organs, 331.
Eugenics, 447-454.
Eustachian tube, 250, 424.
Evaporation, 32, 398.
Evolution, 455.
Excretion, 3, 9, 209; insect, 51; crayfish, 68; fishes, 75; frog, 86; Paramecium, 133; Amœba, 135; man, 224, 236; plants, 293.
Excurrent, 394.
Exercise, 148, 222, 233, 234, 265.
Exoskeleton, 49.
Experiments, 11, 13, 14, 17, 19, 28, 30, 31, 33, 145, 152, 163, 166, 206, 217, 231, 254, 255, 256, 306, 307, 314, 324, 325, 355-358.
Experimental stations, 439.
Expiration, 157.
Extensor muscles, 147.
Eye, 46, 49, 66, 82, 83, 93, 248, 249, 414.
Eye, care of, 250, 286, 423.

F

Fabricius, 460.
Facets, 49.
Factory conditions, 421.
Fainting, 284.
Family, Edwards, 449; Jukes, 451; Kallikak, 451; Nam, 451.
Farsightedness, 249.
Fascicled roots, 324.

FIGURE 8 In the 1925 edition of *Living Things* (fig. 8*b*), the entry for evolution has been removed (cf. fig. 8*a*, showing the 1924 edition). Additional space between "E" and "F" entries allows for the page to end with the same term so that the rest of the index pages did not need to be altered.

Index 477

Embryology, 9.
Emergencies, 279-288.
Emulsion, 219.
Enamel, 222.
Encysted, 135.
Endosperm, 352.
Enemies, of insects, 59; of crayfish, 69; frogs, 90; birds, 99; plant life, 402.
Energy, 23-27, 33, 73, 151, 185, 235.
English sparrow, 97, 98, 99.
Environment, 40, 453.
Enzyme, 84, 135, 183, 211, 214, 215, 218.
Epicotyl, 352.
Epidemic, 172, 268, 411, 414, 420.
Epidermis, 237, 298, 324.
Epiglottis, 158.
Epithelial cells, 236, 237.
Epithelial tissue, 89.
Erect stems, 319.
Erosion, 361, 362, 400.
Esophagus, 74, 83, 211, 213.
Essential organs, 331.
Eugenics, 447-454.
Eustachian tube, 250, 424.
Evaporation, 32, 398.
Excretion, 3, 9, 209; insect, 51; crayfish, 68; fishes, 75; frog, 86; Paramecium, 133; Amœba, 135; man, 224, 236; plants, 293.
Excurrent, 394.
Exercise, 148, 222, 233, 234, 265.
Exoskeleton, 49.
Experiments, 11, 13, 14, 17, 19, 28, 30, 31, 33, 145, 152, 163, 166, 206, 217, 231, 254, 255, 256, 306, 307, 314, 324, 325, 355-358.
Experimental stations, 439.
Expiration, 157.
Extensor muscles, 147.
Eye, 46, 49, 66, 82, 83, 93, 248, 249, 414.
Eye, care of, 250, 286, 423.

F

Fabricius, 460.
Facets, 49.
Factory conditions, 421.
Fainting, 284.
Family, Edwards, 449; Jukes, 451; Kallikak, 451; Nam, 451.
Farsightedness, 249.
Fascicled roots, 324.

FIGURE 8 (*continued*)

One might come away from a reading of the revised *Living Things* with an impression that Darwin's principal importance to the history of biology was his groundbreaking "interesting essay" on earthworms. This bizarre image was reinforced by Allyn and Bacon's 1929 *New General Biology* by Smallwood, Reveley, and Bailey. The book was a revision and expansion of those authors' 1924 *New Biology*. Already in 1924 the *New Biology* used the word *evolution* only in two biographical insets—of Henry Fairfield Osborn and Charles Darwin. The biographical insets—nonpaginated and interspersed throughout the book—were reused in the 1929 revision. The lack of page numbers meant that the plates could be recycled for use in several textbooks, and many of them had also been used in earlier biology textbooks by the same authors. In 1924, Darwin's biography was placed in a chapter on natural selection. In 1929, it was placed after a section entitled "Earthworm and Plant Life."[14] In an apparent attempt to reinforce biology's spiritual nature, that same chapter on earthworms includes a quotation from Longfellow's poem "The Fiftieth Birthday of Agassiz":

> "Come wander with me," she said,
> "Into regions yet untrod;
> And read what is still unread
> In the manuscripts of God."[15]

Louis Agassiz (1807–73), the Harvard zoologist lauded by the poet, was one of the last major scientists in America to die unconvinced of Darwin's theory.

Like Gruenberg's *Biology and Human Life*, the *New General Biology* contained substantial discussion of evolution, even though it did not use the word. The chapter on reptiles begins with a lengthy discussion of dinosaurs and the fossil evidence for their antiquity. The chapter on mammals ("the Rulers of the Earth") discussed the adaptations of homologous structures. The book covers eugenics. It even discusses the origins of new plants and animals quite explicitly:

> Biologists believe that these several kinds have been derived or developed in nature. We are sure that you recall that animals produce more eggs than ever reach maturity and that the number of seeds on one plant is often very great. When these eggs and seeds begin to grow, there is competition between them for food and a place to live. Those that can grow a little faster or are more hardy or more adaptable live; the rest fail to mature or die in infancy. Darwin used the expression "struggle for existence" to describe his conception of this competition with environment and with other animals and plants; Wallace, another English biologist, preferred to use the expression "survival of the fittest" to explain this same condition.[16]

No matter what Darwin or Wallace called it, Allyn and Bacon's editors preferred not to use the word *evolution*. The firm's 1916 biology textbook (by the same authors as the *New General Biology*) explicitly taught the "theory of evolution."[17] Like Ginn, Allyn and Bacon took deliberate steps to respond to antievolutionism.

In the cases of Ginn's *Biology and Human Life* and Macmillan's *Biology and Human Welfare*, the decision to avoid the term evolution was decided very early in the process of writing the manuscript, allowing much of the evolutionary content to be included without "objectionable phraseology." With Iroquois's *Living Things*, the changes were made ex post facto and were constrained by the economics of printmaking.

The most awkward situations occurred when a decision to change the text in response to antievolutionism came late in the production of a new book but before its publication. The ABC's and Hunter's creation of the *New Civic Biology* suffered from this. William H. Atwood's 1927 *Biology* seems to have been revised under similar circumstances. The book reused much of his 1922 *Civic and Economic Biology*, which had devoted several chapters to evolution. Despite explicit discussion of the common ancestry of different species, however, there was no mention of evolution at all in the 1927 book. The word did not appear anywhere in the main text or the index. However, it did occur in the glossary:

> *Evolution* (Lat. *E* + *volvere*, to roll) (ev o lû´ shun): The progress of life by descent from the simple to the complex.[18]

The 1922 textbook had no glossary. The glossary was made from scratch for the 1927 *Biology*. At one time, apparently, the book's manuscript contained the word *evolution*. Somewhere in the editing process—perhaps after a reading of the *New Civic Biology* or other competing books—a decision was reached to remove all mention of it. Someone—presumably an editor—overlooked the glossary when the removal was carried out. This suggests that the decision to remove the word both came late in the production process and was hastily executed. The glossary entry is a unique instance of the often-invisible editorial process being revealed in the final published text.

ALFRED KINSEY AND THE BEGINNING OF THE NEXT GENERATION OF BIOLOGY TEXTBOOKS

Not all the textbooks published just before or soon after the Scopes trial avoided discussion of evolution. Alfred Kinsey's 1926 *Introduction to Biology* made a compelling case for evolution and even went so far as to mock those who claimed to disbelieve it:

> Realizing all of this, wouldn't it seem strange to you if some one should ask you if you really believe that organisms change their characters and become new things?

> But the scientific word for change is *evolution,* and there are some people who think they don't believe in evolution. The man who says so may own a new breed of dog; he wears clothing made of new kinds of cotton, or wool from an improved variety of sheep; he eats bread made of an improved wheat, and buys kinds of corn and potatoes and fruits and things that were unknown to his grandparents; he grows a cut-leaved, newly developed kind of beech or birch or maple on his lawn, and new varieties of roses and chrysanthemums in his garden; and he may smoke a cigar made of a very recently improved tobacco. When he says he doesn't believe in evolution, I wonder what he means![19]

Kinsey's textbook focused primarily on the artificial selection of plant and animal breeds and did not discuss the evolution of humans, but it does use the word *evolution* and, as seen in the passage quoted above, goes out of its way to call attention to the political movement against it. (The word *evolution* was not, however, included in the index.) Kinsey's was one of the few books that did not adhere to what might be called the revision strategy of the period. Of course, the *Introduction to Biology* was not a revision and, as such, was one of the few high school biology textbooks published in the late 1920s that was not based on an earlier book by the same author. Kinsey was also trained as a zoologist, and, though he may have been influenced by Benjamin Gruenberg (who wrote much more about Kinsey and sex education in later decades), he was not a part of the network of biology educators who had played such a central role in the creation of civic biology and its propagation through textbooks in the 1910s and 1920s. Both Kinsey as an author and J. B. Lippincott as a publisher were new to the high school biology market.

It is not clear how important Kinsey's response to antievolutionism was to the success of his textbook. Most of its sales came in the 1930s, in adoptions several years after the Scopes trial and after the issuance of most of the posttrial revised textbooks. Many of these sales were in Texas, where Governor Miriam "Ma" Ferguson had the subject of evolution physically cut out from the state's adopted biology textbooks.[20] It was probably more important that Kinsey's book was much less expensive than its competitors. Its retail price in Kentucky was $1.22, less than the next cheapest book (Clement's *Living Things* at $1.38), and much less than the other five textbooks adopted (whose prices averaged $1.53). In some cases, Kinsey agreed to cut his royalty percentage as he and Lippincott set out to undercut the competition.[21] This may explain why his book sold as well as it did—especially after the economic collapse of 1929.

The *Introduction to Biology* also represented a shift away from the civic biology model of the 1910s and 1920s. Kinsey focused more on ecological principles than urban and industrial applications of the life sciences. Some reviews of the *Introduction to Biology* recognized it as significantly different from the civic biology mode that had characterized high school biology since 1914—a mode that was retained in revised textbooks after the Scopes trial. In the words of one southern pedagogy expert, the University of North Carolina professor (and author of the "Science Column" for the *High School Journal*) Carleton E. Preston, Kinsey's book "was probably the best high school biology text of its particular period from the standpoint of the nature-lover."[22] Preston also praised the book for not "following the customary 'human welfare, health and wealth' trend of biology texts for high schools."[23] A review in the *Peabody Journal of Education* (written by Jesse M. Shaver) was far more critical, calling it "hardly up to the standard set by Moon, Hunter, and others."[24]

Kinsey's and Lippincott's departure from the paradigm of civic biology may have allowed it to reach a market that wanted neither civic biology nor the antiquated biology-zoology divide. This novelty may explain why their book experienced the success that it did. Though by no means a top seller in the biology marketplace, it sold close to 100,000 copies nationwide before being supplanted by an updated edition in 1933. In Kentucky, the textbook was adopted in 1930 along with nearly all the major post-Scopes revised civic biology textbooks (Hunter's *New Civic Biology*, Smallwood, Reveley, and Bailey's *New General Biology*, Clement's *Living Things*, Gruenberg's *Biology and Human Life*, and Peabody and Hunt's *Biology and Human Welfare* as well as *Modern Biology*, by Harry D. Waggoner).[25] Texas adopted Kinsey's book along with several others in 1932; however, it was outsold by each of the other textbooks the state approved. (The two largest sellers in the state were Clement's *Living Things* and Smallwood, Reveley, and Bailey's *New General Biology*.)[26] It was also adopted in a few other southern states, including North Carolina (where Preston was located), but not Tennessee (where Shaver was).[27]

Kinsey's book is less of a post-Scopes textbook and more of a pioneer of the next generation of textbooks after civic biology. Its sales show that not all the developments in biology education in the second half of the 1920s were reactions to the Scopes trial or to the antievolution movement. Its greater ecological focus and reduced emphasis on urban applications reflect a changing set of social priorities in the teaching of biology that became more influential in the 1930s with the textbooks of Ella Thea Smith and other new writers.[28] These books focused on bringing a more integrated approach to nature that treated humanity as part of the natural environment. With greater discussion of applications of biology to agriculture, they also heralded the end of an urban-rural split in high school biology teaching. Although Kinsey's book was published soon after the Scopes trial, most of its sales came in the 1930s, just before the publication of Smith's first textbook in 1932. (An updated *New Introduction to Biology* by Kinsey was published in 1933.) Perhaps this represented another strategy to cope with antievolutionism—to do away with the civic biology approach that had in many ways instigated the school antievolution movement. It is in those books that had earlier versions before the Scopes trial that the way in which publishers responded directly to the antievolution movement can best be seen. To really assess that reaction, it is sufficient to look no later than the early 1930s. The best sense of a revision strategy is revealed by comparing of Tennessee's textbook adoptions in 1925 to the adoption that took place at the start of the next state cycle in 1931.

A TACIT REVISION STRATEGY

It would be hard to describe the evolution of biology textbooks after the Scopes trial as the result of a deliberate strategy. There was no collusion on the part of textbook publishers to uniformly avoid the word *evolution*. Most textbook authors were in communication with one another, but rarely did they exercise much control over this aspect of their texts. George Hunter spoke to Benjamin Gruenberg while the ABC was "emasculating" his *New Civic Biology*. Gruenberg had had a much less confrontational relationship with Charles Thurber and the other editors at Ginn. Hunter and Gruenberg had rather different responses to what their publishers did, but neither of them had cut "evolution" out of their own books.

The perceived success of one book spawned imitations, but these perceptions reinforced themselves. News of Peabody and Hunt's adoption in Tennessee was accompanied by members of the state textbook commission explicitly stating that the book did not contain evolution and the superintendent of the high school where Scopes taught saying that the new book did not contain any evolutionary trees. This gives some impression as to what it meant for a book not to contain evolution. It would seem that a book like Gruenberg's—which had images of fossils described as millions of years old—should not have been acceptable, but, just before Gruenberg declined an invitation to testify for Scopes's defense, it was added to Tennessee's list. Clement's book, hastily (and even sloppily) revised only to remove the word *evolution*, was accepted in Tennessee by October.

While the ABC fell behind its competitors in biology during the long revision of the *New Civic Biology*, Ginn grew more concerned about some of its smaller competitors who had managed to defeat it in major adoption battles despite the quality of its books. George A. Plimpton, the company president, singled out biology as a field in which Ginn needed to improve its sales strategy in an address to the firm in January 1926:

> The great sellers in the high schools of course are the first-year books and in this respect we are not up to our competitors. Take, for instance, biology: there are two books in New York State which have the market—Smallwood and Bailey by Allyn and Bacon, and Clement's by Southworth. Now the point I want to emphasize is this: when we publish a good book, then some one ought to take steps to bring out a revised edition knowing full well that a competitor will bring out a book based on its corrections.[29]

The problem was not just in New York. *Biology and Human Life* had spearheaded the antievolutionary revision strategy *before the Scopes trial even took place*. Yet other publishers, especially Iroquois, were copying Ginn's techniques and winning sales at that firm's expense.

Successful adoptions were interpreted by other companies' salesmen and editors as clues as to what the audience *expected* a biology book to be. As books that very clearly taught evolutionary *concepts* without using the *word* became more and more successful, textbook publishers perceived that removing the word was sufficient for a book to meet local criteria about not teaching evolution. This was not an explicit policy, and not all textbook publishers did this. In part, those that did were influenced by other publishers. But the ABC's George Benton was also influenced by other factors, especially the reports of outside readers.

The idea that the *word* "evolution" was a problem seems to have first originated from outside the textbook publishing world entirely. Thurber cited outside criticism of Gruenberg's textbooks stating that removing the word would be sufficient for antievolutionists. McHenry Rhoads and Jessie M. Shaver expressed similar sentiments in response to Hunter's manuscript. The *Peabody Journal* editorial decrying the Tennessee law suggested something similar. In these cases, the idea that removing the word would be a sufficient strategy came from people who were not themselves antievolutionists. Thurber's correspondence with Bryan may be the only exception, but Bryan did not fully endorse such a misleading revision (nor did Thurber at that time). These other experts held positions of leadership in education in places where school antievolutionism was widespread, which gave them a position of authority from which to explain how best to create a text that would be acceptable to antievolutionary communities. Textbook publishers expected these experts to know and understand the communities they worked in, even when being asked to characterize a position they personally opposed. But these same experts saw Tennessee's antievolutionism as a symbolic protest that "the friends of education swallowed . . . rather than lose the slight majority they could barely command to pass the really crucial education legislation."[30] It may have been that the suggestion to remove the word *evolution* was seen as an equally symbolic response to a symbolic protest, a gesture that school antievolutionists would accept.

If the selection of *Biology and Human Welfare* and *Biology and Human Life* in Tennessee immediately before the Scopes trial was any indication, this understanding of how textbook adopters thought of antievolutionism may have been largely accurate. However, one effect of the Scopes trial was the reinterpretation of school antievolutionism as inherently religious. It altered how antievolutionism was perceived by its opponents and how communities of antievolutionists perceived the issues associated with their opposition. It was not apparent that the same strategies would succeed after the trial.

To some extent, books published after the trial were shaped by perceptions of what antievolutionism meant to the communities that regulated textbooks. But they were also revised with firm attention to the amount of time and money that revisions would take and with reticence on the part of some editors to compromise the quality of their books' content. The removal of the word *evolution* was in part a reflection of the way antievolutionists were seen by their opponents. But it was also the result of intellectual and economic compromises. The decision to include passages emphasizing the moral sensibilities of people, the beauty of nature, and even Darwin's personal religious bearing also shows other ways in which the publishers understood their audience's views of antievolutionism and the science and religion framework that affected those views. The ultimate success of these books means, not that they accurately reflected the real concerns of an antievolutionary society, but that textbook producers were able to sell their books as *responses* to the trial.

READING AND SELLING TEXTBOOKS WITHOUT "EVOLUTION"

It would not have been enough for textbook publishers to decide that the best strategy for high school biology textbooks was removing the word *evolution* and including passages showing religion favorably. For the books to be successful, sales forces would once again have to sell a new kind of textbook and convince adopters that their textbooks would not teach anything objectionable to their children. Jessie Shaver's review stating that Hunter's *New Civic Biology* did not contain evolution was a great help to the ABC. A clipping of this review from one of the most respected education journals in the South was an influential weapon in a salesman's arsenal. It affirmed to readers that, as understood by an expert in education, the book did not teach evolution. At the same time, the letters that Charles Thurber exchanged with William Jennings Bryan in 1923 were circulated among the directors of Ginn in September 1925. What had been suggested as a method of writing or revising textbooks had now become instructions for how to read and market their new textbook in the months following the Scopes trial.[31]

Despite the rise in regulation, textbooks were sold in the late 1920s and early 1930s in a way that was still similar to how they had been sold in the late nineteenth century: they were pitched and adopted with little direct reference to their content. Textbook regulation did little to affect the role of content in adoptions. When Tennessee passed a revised textbook law in 1927, the state textbook commission was obliged not to select books containing a partisan bias, but was instructed to give substantial consideration to the cost and physical durability and quality of the book. State-level regulation had taken away some of the element of personal relationship between textbook salesmen and local adopters and significantly curtailed corruption. (The Tennessee law also forbade anyone who had ever been employed or given gifts by a publisher from serving on the state commission. The $100 that W. T. H. Howe paid Jesse Shaver for his report on the *New Civic Biology* manuscript might have disqualified him.) Nonetheless, textbook salesmanship emphasized the book's physical quality and the publisher's ability to meet demand. Shaver's review would have been a sufficient testimonial to assure adopters that the book was acceptable with regard to its nontreatment of evolution.

Reviews like Shaver's essentially proposed an interpretation of the book in question. In this case, that interpretation was the following: a book not containing the word *evolution* does not contain evolution, even if it includes discussion of heredity and variation and the development of plants, animals, and humans. Salesmen not only used these interpretations; they offered their own to suit the desires of potential adopters. ABC sales agents reported that, before the *New Civic Biology* was available, other companies were using news of the Scopes trial to attack the old *Civic Biology*. Even though, technically, the old book did not contain anything strictly prohibited by the Tennessee law, rival salesmen had effectively sold an interpretation of the textbook that rendered it unacceptable.

It was even possible for textbook salesmen to sell completely opposite interpretations of the same book to different markets. In the October, just after the Scopes trial, James Peabody sent Henry Fairfield Osborn a copy of some of the promotional materials that Macmillan was using to sell *Biology and Human Welfare*. In addition to pamphlets for distribution, this included a typewritten outline of arguments and a collection of quotations from the book meant to be used to convince regulators that the book would meet their needs when it came to the evolution question. The first thing that the sales memorandum notes is that *Biology and Human Welfare* was "adopted as 'safe' for use in high schools by the Tennessee Textbook Commission" and "held up by the California officials because it teaches 'evolution.'"[32] The question of whether the book taught evolution depended less on the text itself than on the way in which the book was sold.

Textbook salesmen were in a unique position to market interpretations along with the actual textbooks. The people responsible for selecting and using textbooks already relied on salesmen and the textbooks themselves to inform them of new developments in pedagogy and curricula. Where teachers were not well trained, it was often expected that they would rely on the textbooks to dictate the content of their courses. This dependence on textbook salesmen and their products was not ideal from the point of view of education experts, who had long decried the power textbook publishers held in American education. They saw the reliance on textbook salesmen as proof that more teacher training was imperative. However, this dependence continued unabated in many places after the Scopes trial, prompting the University of Iowa professor of education E. J. Ashbaugh to lament in 1926:

> Textbooks are used by the children and teachers because they have been adopted. They are adopted by lay boards of education, sometimes only on the advice of the sales agent. Sometimes the adoption follows the recommendation of the superintendent, a group of principals, or a committee of teachers. Rarely, however, has any individual or group been able to go back to the claims of the author or the agent and verify these claims. In general these statements must be taken on faith, faith that these claims contain the whole truth and nothing but the truth. This again places a heavy responsibility upon the author of a text.[33]

While Ashbaugh perhaps attributed too much responsibility to authors and not enough to others involved in textbook creation, he did observe the role sales agents still played in the 1920s in getting their textbooks into schools and the role that those textbooks played once they were in the classroom. How they were used and read determined much of what made a particular interpretation of a text possible.

In what way were these biology textbooks acceptable in states like Tennessee and Arkansas, where antievolution was the law? How could a book like Gruenberg's *Biology and Human Life,* which teaches about Darwin's theory of natural selection giving rise to new variations among species and describes fossils as millions of years old; or Smallwood, Reveley, and Bailey's *New General Biology,* which instructs students to compare the bone structures of moles and bats with their own arms, not be considered to be teaching evolution?

In 1931, Tennessee made its first post-Scopes textbook adoption. It chose *Biology and Human Life, New General Biology,* Clement's *Living Things,* and—most surprisingly—Hunter's *New Civic Biology.* In effect, the state endorsed the strategies of word removal and favorable mentions of religion and did so with the presumption that it was selecting books that would not teach evolution. But, even if members of these adoption boards did attempt to regulate the content of the textbooks, the most cursory glance at the books would confirm their acceptability. The word *evolution* is absent from the indices, glossaries, tables of contents, and the main texts (with only some ancillary mentions of the word in *New General Biology*—probably the least used of the textbooks). If the first step in checking a book's content is to look for the word *evolution* or to look for an evolutionary illustration, the books would survive. (It is most telling that a book that was truly devoid of discussion of evolution but whose glossary included the word by mistake was *not* adopted.)

Members of textbook adoption committees were not experts in biology and quite likely unaware of or unconcerned with details of evolutionary theory. They were not appointed to adopt only biology textbooks or even only science textbooks. With a large number of options to chose from in every topic, would a committee member without scientific expertise, already assured that the biology textbooks under consideration did not contain evolution, need to judge whether discussions of "heredity and variation" still taught something unacceptable? Even if a member of the textbook commission was less than fully satisfied with this word-removal strategy, the alternatives were not books that taught less evolution but books like Kinsey's that taught it aggressively. The authority with which the books were described as not containing evolution, the lack of textbooks that were any less evolutionary than these, and the existence of a plausible interpretation of the books as nonevolutionary made them acceptable. It is likely that most textbook adopters relied on salesmen, on textbook reviews and expert opinions, and on the endorsement (via adoption) of a textbook by other states with similar views.

Ironically, these textbooks were published in the wake of a trial that was explicitly and very self-reflectively about the nature of literal interpretation. That readers who already self-identified as literalists constituted a significant part of their perceived audience may have affected how publishers assumed the textbooks would be read.

CHAPTER EIGHT

Losing the Word: Measuring the Impact of Scopes

The debate over what effects the Scopes trial had on biology textbooks is long-standing. To some extent, that debate began even before the trial itself as newspapers reported that textbook publishers were feeling pressured by the upcoming court case (a claim textbook editors denied).[1] Starting with studies in the 1970s, there have been attempts to quantify the contents of textbooks and to draw inferences about the impact of the Scopes trial from these assessments. In 1974, Judith V. Grabiner and Peter D. Miller concluded: "The teaching content of evolution in the high schools—as judged by the content of the average high school biology textbooks—*declined* after the Scopes trial."[2] Gerald Skoog followed this up in 1979 by observing that, in the early 1920s, "authors discussed evolution openly and attested to its validity" but that, "after 1925, statements of this nature were infrequent until the 1960s."[3]

Both these studies have been very influential, not just for understanding the history of biology textbooks, but in shaping interpretations of the Scopes trial's importance. They have also had an impact on recent debates over the teaching of evolution as other scholars have drawn on their methods (with refinements) to evaluate the content of more recent biology textbooks. Later studies have debated whether the best way to measure the evolutionary content of a textbook is "to measure the percentage of a text devoted to evolution" or to conduct a "simple, numerical word count" of evolutionary keywords.[4] These quantitative techniques may have improved on Grabiner and Miller's original study, which was more anecdotal and directly compared specific books like Hunter's *Civic Biology* and *New Civic Biology*. But the newer studies did not question the primary assumption that *the Scopes trial* was responsible for apparent changes between the early and the late 1920s. Even in the twenty-first century, there is debate over the extent of textbook differences, but no questioning the premise that such changes were caused by the Scopes trial. These studies have presupposed that the effects of the trial can be seen in the textbooks' content, with little attention paid to how these books were actually sold, used, and read.

Shortly after Grabiner and Miller's article was published, the paleontologist George Gaylord Simpson wrote that their study was "an instance of mistaking effect for cause."[5] In fact, both the causes and the effects require more careful attention. The antievolution movement was already affecting textbooks *before* the Scopes trial. Charles Thurber's consultation with William Jennings Bryan took place well before the Tennessee law was passed. Peabody and Hunt's textbook, which claimed not to teach evolution, was adopted in Tennessee a month before the Scopes trial began. George Hunter had even been working on a new edition of the *Civic Biology* before the trial. The changes made to Hunter's book reflect both the influence of the trial itself and the success of strategies that had their beginnings before the trial. It is a better and more meaningful question to ask about the impact of the *antievolution movement* on textbooks, rather than isolating the Scopes trial as the sole cause. One must not lose sight of biology textbooks' impact on the antievolution movement and see them merely as affected by the controversies. Influence goes both ways.

But there are no easy answers to even this improved question. Many of the textbooks published either shortly before or years after the trial avoided the use of the word *evolution*. By a word-counting measure, these books might be considered to have removed or diminished their discussion of the topic. But many of them replaced the word with an equivalent term or phrase and retained much of the same biological content. A more thematic measure of evolution that includes other keywords might show less of a decrease.

But neither of these criteria measures how these textbooks were marketed and read. Historians of publishing have long criticized the idea that a book can control the way in which readers engage it. In the case of textbooks—whose authority is reinforced by a system of state regulation and publisher marketing—there is a greater sense that readers (students) can have texts' meanings prescribed for them, though the books themselves do not do so.[6] Tennessee's adoption of the *New Civic Biology* while evolution was prohibited, reinforced by salesmen-procured endorsements stating that the book did not contain evolution, meant that most people in Tennessee thought that the book was nonevolutionary. But the *New Civic Biology* could have been interpreted as teaching evolution, and in states where it was less problematic to do so it likely was.

More than the number of evolutionary keywords, the actual influence the Scopes trial and the antievolution movement had on textbooks can be best understood in terms of how those textbooks were taught and interpreted. Evolutionary theories of mind were a key part of the intellectual framework that had led to the curriculum of civic biology, and that model of educational psychology spoke directly to how students thought, read, and learned. The *Civic Biology* was never concerned with the teaching of the evolutionary development of life as just an intellectual curiosity. Evolution and other key concepts of biology were brought into science education as part of the new understanding of schooling as a form of social development. Scientific concepts could be applied to an urbanizing and modernizing life. This view of education extended to changing the way young minds interacted with textbooks. Students were being taught to think inductively and to eschew the rote presentation of facts. Back in 1916, John Dewey lamented the "isolation of science from significant experience": "The pupil learns symbols without the key to their meaning."[7] The most important sense in which biology (and other) textbooks were evolutionary was in their pedagogical commitment to the development of student minds, teaching them to connect symbols to their deeper meanings through the relationship of scientific theories to real-world experiences.

These post-Scopes biology textbooks, presented in a context in which certain readings were prescribed, lost sight of this evolutionary aim of education. Reading a book that taught the heredity and development of species without using the word *evolution* and being compelled to interpret these books as not teaching anything that the antievolution law prohibited encouraged a reading that relied on literal interpretation. This was a return to rote: an education that denied student abstraction that might have led to interpretations of the books as evolutionary. This form of pedagogy wholly undermined the grandest ambitions of civic biology and the reform of science education of the first two decades of the twentieth century: to apply the principles of the evolution of the child mind to bring about social progress.

By the late 1920s, evolutionary theories of mind were beginning to lose influence among American psychologists.[8] Perhaps this helps explain why there was not a greater outcry. The change in textbook reading practices was obscured by the more superficial ways in which biology textbooks lost evolution. But the idea that those textbooks' coverage of evolution disappeared in the late 1920s helped pave the way for a story about its resurrection in the 1960s.

THE RUIN AND RESTORATION OF EVOLUTION

Regardless of how the concept is measured, evolution never really disappeared from biology textbooks. In his criticism of Grabiner and Miller, Simpson pointed out that textbooks by Ella Thea Smith published in the 1930s and 1940s *did* discuss evolution.[9] There was also Alfred Kinsey's textbook, whose sales were not as bad as Grabiner and Miller alleged.[10] Nonetheless, the prevailing story of evolution in American schools is that it disappeared after 1925 and remained absent until the 1960s.[11]

Ronald P. Ladoucer has argued that this narrative is largely a "myth, created first as part of a public relations effort by the Biological Science Curriculum Study (BSCS) to differentiate, defend, and promote its work, and later as part of an attempt by scholars to sound a warning concerning the rise of the religious right."[12] Funded by a National Science Foundation grant in 1958, the BSCS was one of several initiatives meant to "transform" science education in the United States after World War II.[13] These initiatives proceeded in ways similar to those of the early twentieth century, when the National Education Association committees had studied and suggested education reform—including the development of science education tied to real-world experiences like civic biology and the incorporation of the evolutionary pedagogy emerging from thinking like Herbert Spencer's, G. Stanley Hall's, and John Dewey's. John L. Rudolph has shown that, despite the popular conception that this reform of science education was a response to the Soviet launch of the first man-made satellite in 1957, many of the initiatives had begun months or years before America's first "Sputnik moment."[14] The BSCS resulted in the development of new biology curricula and new textbooks whose incorporation into American high schools was hailed as part of a new era in American science education. Along with similar programs in other sciences, it represented a greater level of federal involvement in science education than previously seen in the United States. BSCS supporters also invoked the specter of the Cold War to emphasize the role of science education in recruiting and training scientists who would allow the United States to compete with the Soviet Union. This represented a major shift in the aims of science education. In the era of civic biology, science educators were concerned less with producing new scientists than with instilling a scientific ethos in all students, future scientists and nonscientists alike.[15]

Public attention to the teaching of evolution in the United States intensified in the late 1950s and early 1960s. The popularity of the 1955 play *Inherit the Wind* and the subsequent 1960 film, which presented a fictionalized account of the Scopes trial, reinforced many of the mythic distortions of what happened in Dayton.[16] The year 1959 was also the centenary of the publication of the *Origin of Species*, and the conferences and public events commemorating the impact of Darwin and his thought also resulted in renewed popular discussion of Darwinism.[17] These popularizing effects coincided with the pedagogical efforts of the BSCS.

Ironically, one of the people most responsible for the BSCS-era popularization of the idea that evolution disappeared was George Gaylord Simpson. In 1960, he wrote an influential article, "One Hundred Years without Darwin Are Enough," in which he argued that American high schools were not in fact teaching evolution. But he claimed that it was not antievolution *laws* that were responsible for this. "Laws against evolution are still nominally in effect," he noted. But they were not really enforced. Antievolutionism had won out largely because teachers and textbooks avoided the subject. Fear of controversy, not of the laws themselves, had led teachers and textbook publishers into self-censorship.[18]

This changed with the advent of the BSCS textbooks (first published in 1963) and other biology textbooks whose publishers rushed to copy their techniques. The BSCS textbooks emphatically presented evolution and reawakened the legal and political attention to antievolutionism.[19] In 1965, Arkansas (which in 1928 had passed an antievolution law prohibiting both the teaching of evolution and the use of textbooks containing evolution) adopted a biology textbook that taught evolution. This led to a test case similar to Scopes. Rather than be prosecuted for criminal violation of the antievolution law, the Arkansas schoolteacher Susan Epperson preemptively sued the state for putting her in an impossible position, one that violated her due process rights. (If she did not teach the textbook, she would violate the law requiring teachers to use adopted texts; but, if she did use the textbook, she would violate the antievolution law.)[20]

Previously, in 1947, in another case concerning the rights of states to control their schools, the U.S. Supreme Court had ruled that the First Amendment prohibition against the establishment of religion applied to the states as well as to the federal government.[21] On the basis of this precedent, the Court ruled in Susan Epperson's favor and declared that state prohibitions of the teaching of evolution were unconstitutional. Perhaps anticipating this, the state of Tennessee repealed its antievolution law in 1967, a year before *Epperson v. Arkansas* was decided.

Reaction to the *Epperson* ruling helped popularize an effort by antievolutionists to verify their creationist conclusions under the rubric of "science." As Ronald L. Numbers points out, "the transmogrification of creationism from religion to science took place in direct response to" textbook adoption controversies in California in the early 1970s, "which encouraged creationists to believe that they could squeeze into science classrooms simply by shedding superfluous biblical weight."[22] "Creation science" or "scientific creationism" continued a paradigm of "science and religion" that came into view during the Scopes trial, which saw the definitions of both *religion* and *science* subject to debate. Though the Scopes debate centered more on whether ideas such as biblical literalism and theistic evolution could count as religion, the post-*Epperson* reaction of antievolutionists was to argue that their version of creation could count as science.

The antievolutionism of creationists and creation science proponents was very different from the antievolutionism of the 1920s. Unlike Bryan, they espoused the young age of the earth and insisted on the occurrence of a worldwide flood to account for observations such as those of fossils layered in various geological strata. These young earth creationists held views more similar to Darrow's caricature of Bryan than to those of most antievolutionists of the Scopes era.[23] The version of creation that their "science" was designed to confirm was the literal interpretation of Genesis.

After *Epperson,* opponents of evolution proposed laws that required giving "balanced treatment" to evolution and creation science. It is debatable whether this was done to actually promote creation science or more simply to revive antievolution laws by other means (by discouraging evolutionists from teaching evolution so that they would not have to teach creation science). But the legal fight that emerged over these laws once again focused on redefining *science and religion*. The 1981 federal trial *McLean v. Arkansas Board of Education* ruled that Arkansas's newly enacted balanced-treatment law was unconstitutional because it represented an establishment of religion. The court effectively determined that creation science was religious in large part because it was not scientific. This legal reasoning enshrined as precedent the notion that science and religion were in opposition and mutually exclusive—a view of conflict that the Scopes trial helped make prevalent. This legal precedent has also led subsequent antievolutionary theories or evolution "alternatives" to be subjected to a de facto test of whether they merit inclusion in the category *science*.[24] The legal and political debate over evolution continues to rely on the science and religion legacy of the Scopes trial. It also relies on the trial's perceived legacy to textbooks.

SCOPES MYTHS AS PRECEDENTS

The idea that evolution in biology textbooks "declined" after 1925 has been central to ongoing legal debates over the teaching of evolution and its "alternatives," "criticisms," or "strengths and weaknesses." This mutually reinforcing connection between the effects of the Scopes trial and biology education came to a head in 2005 with the verdict of the *Kitzmiller et al. v. Dover Area School District* trial in Pennsylvania. A school district that encouraged the study of "intelligent design" (ID) was sued by the parents of some high school students. Judge John E. Jones III's ruling that ID was not a scientific theory and could not be taught in public schools drew largely on the fact that there was a historical pedigree connecting ID to other forms of creationism and, through that, to earlier forms of religious antievolutionism.[25] That historical lineage was established in large part by an examination of the ID textbook recommended by the Dover school district, entitled *Of Pandas and People*.[26] The early drafts of this textbook had been written prior to a 1987 Supreme Court ruling that upheld the opinion in the *McLean* case declaring unconstitutional the teaching of creationism or creation science in public schools. In the Dover trial, the court found that "cognates of the word creation (creationism and creationist), which appeared approximately 150 times [before 1987] were deliberately and systematically replaced with the phrase ID."[27] The editors of *Pandas* had drawn on the same strategy as those post-Scopes biology textbooks that had replaced only the word *evolution*.

Rejecting the argument that the "intelligent design" taught in *Pandas* was not creationism, Judge Jones proved himself to be a different kind of text reader than the 1931 Tennessee state textbook commission. He reasoned that the interpretation of the text put forward by its publishers was not consistent with how students would actually read. Students would interpret the book as having a religious meaning despite the absence of the explicitly religious term *creation* because they would be familiar with the religious pedigree of antievolution movements.

Opponents of antievolutionism characterized ID as equivalent to creationism, in part because the authors of *Pandas* had not changed much besides substituting the phrase *intelligent design* for the word *creationism*. They did not change the arguments or the evidence they used. Nonetheless, ID advocates claim that intelligent design is science. They point to the fact that they make no direct statements about the truth of the Bible as proof that what they are doing is not religion.[28] This sense of *religion* as Bible-oriented comes directly from the understanding of *science and religion* that emerged in Dayton. Bryan's rejection of theistic evolution as a religious position aligned antievolutionism with views of religion that were more connected to the Bible and to revelation than to philosophical or natural religion. While Bryan did not claim that evolution was against the Bible because it contradicted the literal interpretation of Genesis, he did claim that the irreligious nature of evolution undermined belief in the truth of the Bible. ID advocates invoke the fact that they are not biblical literalists to support their claim that ID is not religion and make tacit use of the logic that, if it is not religion, it may be science.

In Dover, a member of the school board read students a prepared statement (after the school's biology teachers refused to do so) encouraging them "to keep an open mind." That statement also referred to *Of Pandas and People* as a "reference book." This is another example of how school regulators attempt to shape the perceived authority of textbooks, the classroom, and course content. The reminder that "Pennsylvania Academic Standards require students to learn about Darwin's Theory of Evolution and eventually to take a standardized test of which evolution is a part" shows the local school board attempting to undermine the authority of the state in dictating education content.[29] That is, the implication that evolution is taught only because of the standardized test encourages students to learn no more about the subject than is necessary to conform to the law, to study it in as rote a fashion as possible. Although the language of open-mindedness suggests that the school board encouraged critical evaluation, the rest of the language suggests the greater credibility of intelligent design (described as an "explanation" rather than as a "theory . . . not a fact," like evolution is).[30]

This statement in Dover was one of several examples in recent decades where state and local authorities have tried to dictate modes of reading and interpreting biology textbooks that undermine their evolutionary content. Several jurisdictions have required stickers to be placed in the front of their biology textbooks containing warnings about the books' evolutionary content. While the language employed has varied somewhat from place to place, the stickers frequently invoke rhetoric that goes back to William Jennings Bryan's distinction that evolution is a "theory, not a fact." In these cases, as with the statement read aloud to Dover, Pennsylvania, students, the structure of textbook adoption and regulation was not embellishing the authority of the textbooks being presented to students but systematically undermining it by presenting a competing authority.[31] The battle of science and religion became a battle of Pennsylvania versus Dover or Alabama versus previous federal court decisions.[32] Shortly after the *Kitzmiller* verdict, a federal appeals court upheld a district court ruling that the insertion of stickers like these in biology textbooks in Cobb County, Georgia, was unconstitutional.[33]

The legacy of the Scopes trial and its influence on current debates is a complex one that should not leave any party satisfied. Textbook reading and use has been stripped of some of the inductive, nonrote learning that evolutionary educators like John Dewey, George W. Hunter, and Otis Caldwell thought were the most important improvements to science education. But, if widespread acceptance of a less inductive engagement with text represents a loss for evolutionists, it is at best a pyrrhic victory for the antievolutionist. The *kinds* of antievolutionism that have become so widespread are often rooted in the literal interpretation of the Bible that early Fundamentalists had disavowed. In the legal demand that alternatives and criticisms of evolution be fit into the rubric of science, religious antievolutionists have ceded the discussion of their reading practices to their own "scientific" standard bearers. The most prominent spokesmen for ID are those who identify themselves as scientists. This makes it impossible for antievolutionists to engage in fruitful discussion of the religious implications of their various theories. Antievolutionary alternatives are judged, not on their intellectual merits or theological virtues, but on their ability to pass as "science."

Antievolutionism as "scientific" was emphasized by a 2012 Tennessee law declaring "teachers shall be permitted to help students understand . . . the scientific strengths and scientific weaknesses of existing scientific theories covered in the course being taught." Critics panned this as a "money bill."[34]

These are consequences of a kind of thinking about science and religion that has changed little since the Scopes trial. The trial's participants ignored the specific contexts in which their debate emerged and redefined the nature of both science education and antievolutionism. In nearly all the antievolution cases brought to trial in America since 1925, the issues of who controls the schools, what textbooks are used there, how they are read, and on whose authority they are understood have been submerged in a narrative of essential conflict. Participants in these cases, assuming the birthright of the Scopes trial, return to the trope *science and religion*. In doing so, they present their cause as something unchangeable—like a dogma—as if they meant the warfare of it to outlive them.

Acknowledgments

One of the major themes this work explores is the compound and complex nature of authorship and the fact that many people who contribute to a book go unnamed or unacknowledged. Unlike some of the textbook authors I discuss, I have been fortunate that the various individuals and organizations that have helped shape this book have been overwhelmingly supportive. I am indebted to them, not only for making this book much better than it could otherwise have been, but also for making the research and writing experience a positive one.

The research for this book began at the University of Chicago. Adrian Johns and Robert J. Richards were both exceptionally generous with their guidance and expertise, which helped me turn a collection of sources and insights into a coherent project. Ronald L. Numbers has provided additional mentorship and enthusiasm for this project and during my appointment at the University of Wisconsin has been a source of constant advice and encouragement.

The research reflected in this work required several trips to archives across the United States. The Fishbein Center for the History of Science and Medicine at the University of Chicago funded many of those trips. The University of Chicago Humanities Division also provided research funding during the early stages of this work. Additional support came from the Albert Shanker Research Fellowship for Research in Education at Wayne State University, a Princeton University Library Research Grant, a Dibner Fellowship in the History of Science at the Huntington Library, and a National Science Foundation Postdoctoral Award. On many of these trips, I came to rely on the expertise of many librarians and archivists who helped me locate sources and offered many suggestions for material to examine. In particular, I would like to thank Victoria Alfasso, the librarian at DeWitt Clinton High School in the Bronx, for providing me with access to the school's collection of yearbooks; Linda Hall at Williams College, who sent me material concerning the school's science curriculum; Lance Factor at Knox College, who sent me material about George Hunter's time there; and Thomas Black, the former registrar at the University of Chicago, who helped me access transcripts whose information proved crucial.

Several other people volunteered their time and expertise to read parts of this book or to discuss this work with me. The comments and criticisms of Keith Benson, Dawn Biehler, Robert Brain, Constance Clark, Christopher DiTeresi, Donna Drucker, Erika Dyson, Kimberly Hamlin, Marcie Holmes, Ronald P. Ladoucer, Lynn K. Nyhart, John L. Rudolph, Steve Wald, Cecelia A. Watson, William C. Wimsatt, and many other colleagues have all been greatly appreciated. Matthue Roth has been pushing me as a writer since our youth, and Katie Ramos has been a tireless and enthusiastic copyeditor in the manuscript's final stages. I am also grateful to the Martin Marty Center at the University of Chicago Divinity School and the Zygon Center for Religion and Science for giving me the opportunity to workshop parts of this project in a religious studies setting and to the University of Chicago School Mathematics Project for giving me the opportunity to work as a textbook editor for two years.[1]

Thanks are especially due to the family and friends who put up with my strange fascinations with textbook publishing, biology, and the Scopes trial for years. Mike, Ben, Kjell, Emily, Laura, Povl, Ariel, Jena, and Victoria have all been hearing about this project for years and have never been shy about making suggestions and asking questions. My parents, Robert and Cheryll, and my brother Michael have been unfailingly supportive of having a historian in the family. And my partner, Stacey Bhaerman, has been a constant companion and a reminder of what's important, especially when I feel like I don't know much.

Notes

CHAPTER ONE

1. P. L. Harned, "Facts about the State Adoption of Schoolbooks," unpublished draft of memorandum, 5, and Bruce P. Shepard to P. L. Harned, March 21, 1924, folder 5, container 98, Governor Austin Peay Papers, and P. L. Harned, memorandum, n.d., Commissioner of Education Records, 1913–70, Record Group 92, Tennessee State Library and Archives (TSLA), Nashville.

2. Other southern states, along with dates of most recent adoption, are as follows: Mississippi, 1920; West Virginia, 1922; and Alabama, Arkansas, North Carolina, South Carolina, Texas, and Virginia, 1923. P. L. Harned to Austin Peay, n.d., folder 5, container 98, Peay Papers.

3. Austin Peay to P. L. Harned, January 3, 1924, folder 5, container 98, Peay Papers.

4. Lon C. Hill to Austin Peay, February 26, 1924, folder 2, container 44, Peay Papers.

5. For example, Austin Peay to Mrs. Neil Wright, March 26, 1924, folder 2, container 59, Peay Papers.

6. P. L. Harned, *List and Prices of Text Books Adopted in 1919 and Prices on the Same Books from September 1, 1924 to June 30, 1925* (Nashville: State of Tennessee, 1924), 3.

7. P. L. Harned to Austin Peay, October 16, 1924, folder 1, container 59, Peay Papers.

8. Harned, "Facts about the State Adoption of Schoolbooks," 4.

9. Harned, *List and Prices of Text Books*, 3.

10. George W. Hunter, *A Civic Biology, Presented in Problems* (New York: American Book Co., 1914).

11. Edward J. Larson, "Law and Society in the Courtroom: Introducing the Trials of the Century," *University of Missouri–Kansas City Law Review* 68 (2000): 543–48.

12. Ronald L. Numbers, *The Creationists*, expanded ed. (Cambridge, MA: Harvard University Press, 2003), 319–40, and *Darwinism Comes to America* (Cambridge, MA: Harvard University Press, 1998), 76–91.

13. Edward J. Larson, *Summer for the Gods: The Scopes Trial and America's Continuing Debate over Science and Religion* (Cambridge, MA: Harvard University Press, 1997). As implied by Larson's subtitle, the principal subject of his history is science and religion. (The choice of subtitle was the publisher's and not Larson's. Larson nonetheless situates the origins of the Scopes trial in the conflict, or the perception of conflict, between science and religion [see esp. ibid., 11–30].) Earlier histories of the trial also make use of this theme or focus even more directly on the clash between Darrow and Bryan. See, e.g., Leslie H. Allen, *Bryan and Darrow at Dayton: The Record and Documents of the "Bible-Evolution Trial"* (New York: Arthur Lee, 1925).

14. In the Middle Tennessee division from September 1919 to the end of 1923, 5,900 copies

of Hunter's *Civic Biology* were sold at contract price and another 647 at exchange prices. "High School Books Sold and Exchanged from September 1, 1919 to January 1, 1924 by Middle Tennessee Book Depository," memorandum, Commissioner of Education Records, 1913–70, Record Group 92, TSLA.

15. *Official Copy of the Proceedings of the Text Book Commission of Kentucky* (Frankfort: State Text Book Commission of Kentucky, 1924), 17.

16. William Jennings Bryan to C. H. Thurber, December 22, 1923, container 38, William Jennings Bryan Papers, Manuscripts Division, Library of Congress, Washington, DC.

17. "Monkey Talk Hits Book Publisher; Evolution Winner in Nebraska," *Chattanooga Times*, June 5, 1925, 1; "Mr. Scopes Wasn't the First," *New York Times*, June 11, 1925, 18.

18. John T. Scopes and James Presley, *Center of the Storm: Memoirs of John T. Scopes* (New York: Holt, Rinehart & Winston, 1967), 59.

19. William Jennings Bryan to W. B. Marrs [Marr], June 11, 1925, copy, enclosed in Ewing C. Baskette to Clarence Darrow, January 6, 1934, box 2, Clarence Darrow Papers, Manuscripts Division, Library of Congress. This letter was apparently taken from Marr's office in 1931 and copied by Ewing C. Baskette, who sent a copy to Clarence Darrow three years later. Enclosing a copy of this and one other letter, Baskette writes: "I copied as written in the original including the signature. . . . Make what use you please but don't mention Mr. Marr's name as I got this file from his office while I was there about three years ago. I like this kind of stuff. He hasn't missed it so ___." Baskette to Darrow, January 6, 1934. W. B. Marr had organized Bryan's 1924 lectures in Nashville on the subject "Is the Bible True?" Kenneth K. Bailey, "The Enactment of Tennessee's Antievolution Law," *Journal of Southern History* 16, no. 4 (November 1950): 475.

20. Austin M. Peay to Tennessee House, printed in *Journal of Tennessee House, 1925* (Nashville: State of Tennessee, 1925), 744.

21. See, e.g., Allen, *Bryan and Darrow at Dayton*; Larson, *Summer for the Gods*; and Marcel Chotkowski LaFollette, *Reframing Scopes: Journalists, Scientists, and Lost Photographs from the Trial of the Century* (Lawrence: University Press of Kansas, 2008).

22. John T. Moutoux, "Accused Evolution Prof Most Popular Man in Town," *Knoxville News*, May 11, 1925.

23. Larson, *Summer for the Gods*, 24.

24. F. E. Robinson and W. E. Morgan, *Dayton's Cultural Growth, Particularly Agri-Cultural!* (Chattanooga: Andrews Printery, 1925), 2. Though this is the title indicated on the pamphlet, this is frequently cited as *Why Dayton of All Places?* See Larson, *Summer for the Gods*, 400; and "The Scopes Evolution Trial of 1925," http://www.rheacounty.com/scopes.html (accessed July 9, 2012).

25. Robinson and Morgan, *Dayton's Cultural Growth*, 3.

26. Larson, *Summer for the Gods*, 93–95.

27. Robinson and Morgan, *Dayton's Cultural Growth*, 5.

28. H. L. Mencken, "Tennessee in the Frying Pan," *Baltimore Sun*, July 20, 1925.

29. W. E. B. DuBois, "Scopes," *Crisis*, September 1925, 218 (republished as "Dayton Is America," in *The Scopes Trial: A Brief History with Documents*, by Jeffrey P. Moran [Boston: Bedford/St. Martin's, 2002], 182).

30. Jeffrey P. Moran, "Reading Race into the Scopes Trial: African American Elites, Science and Fundamentalism," *Journal of American History* 90, no. 3 (2003): 891–911.

31. Larson, *Summer for the Gods*, 88.

32. *Scopes v. State*, 154 Tenn. 105 (1927).

33. *Kitzmiller et al. v. Dover Area School District*, 400 F. Supp. 2d 707 (M.D. Pa. 2005). Jones quotes from the majority Supreme Court opinion issued in *Epperson v. Arkansas*, 393 U.S. 97 (1968).

34. Larson, *Summer for the Gods*.

35. David B. Wilson, "The Historiography of Science and Religion," in *Science and Religion: A Historical Introduction,* ed. Gary B. Ferngren (Baltimore: Johns Hopkins University Press, 2002), 13–29; Ronald L. Numbers, "Simplifying Complexity: Patterns in the History of Science and Religion," in *Science and Religion: New Historical Perspectives,* ed. Thomas Dixon, Geoffrey Cantor, and Stephen Pumfrey (Cambridge: Cambridge University Press, 2010), 263–82; Geoffrey Cantor, "What Shall We Do with the 'Conflict Thesis?'" in ibid., 283–98.

CHAPTER TWO

1. Adam Lynch, "Grand Hotel: Does the King Edward Have a Glorious Future?" *Jackson Free Press,* August 3, 2005.

2. *Report of the Special Senate Investigating Committee* (Jackson: Mississippi State Senate, 1928), 113, 153–54.

3. Russell Kirk and James McClellan, *The Political Principles of Robert A. Taft* (1967), 2nd ed. (New Brunswick, NJ: Transaction, 2010), 73; Robert L. Fleeger, "Theodore G. Bilbo and the Decline of Public Racism, 1938–1947," *Journal of Mississippi History* 68 (Spring 2006): 19.

4. David G. Sansing, *Making Haste Slowly: The Troubled History of Higher Education in Mississippi* (Jackson: University Press of Mississippi, 1990), 91–110.

5. Clyde L. King, "Legislative Notes and Reviews: Governors' Messages, 1928," *American Political Science Review* 22, no. 3 (August 1928): 646, 638.

6. *Report of the Special Senate Investigating Committee,* 253.

7. Lewie W. Burnett, "Textbook Provisions in the Several States," *Journal of Educational Research* 43, no. 5 (January 1950): 357–66.

8. *Doan v. American Book Company,* 105 F. 772 (7th Cir. 1901).

9. *Ginn v. Apollo Publishing Company,* 215 F. 772 (E.D. Pa. 1914).

10. "Will Down the Trust," *Chicago Daily Tribune,* April 4, 1891, 13.

11. Edwin Ginn, *Are Our Schools in Danger? The Great School-Book Combination* (n.p., 1895), 2. Before appearing in pamphlet form, this essay had previously appeared under the title "Educational *vs.* Political Interests in the School-Book Business" in the *Saturday Evening Post* (June 6, 1891) and *Publishers' Weekly* (June 20, 1891).

12. P. L. Harned, "List of Publishers," memorandum, n.d., folder 2, container 44, Governor Austin Peay Papers, Tennessee State Library and Archives (TSLA), Nashville.

13. John Tebbel, *A History of Book Publishing in the United States,* 3 vols. (New York: R. R. Bowker, 1972–81), 1:551–52.

14. Horace Mann quoted in ibid., 550.

15. Ibid.

16. Joseph Moreau, *Schoolbook Nation: Conflict over American History Textbooks from the Civil War to the Present* (Ann Arbor: University of Michigan Press, 2003), 52–91.

17. Thomas Bonaventure Lawler, *Seventy Years of Textbook Publishing* (Boston: Ginn, 1938), 9, 16.

18. George L. Craik, *The English of Shakespeare: Illustrated in a Philological Commentary on His Julius Caesar* (London: Chapman & Hall, 1857).

19. Lawler, *Seventy Years of Textbook Publishing,* 19; Robert I. Rotberg, *A Leadership for Peace: How Edwin Ginn Tried to Change the World* (Stanford, CA: Stanford University Press, 2007), 22.

20. "Agents' Meeting," memorandum, December 1926, 1, George A. Plimpton Papers, Rare Book and Manuscript Library, Columbia University, New York.

21. Mauck Brammer, *Our Heritage and History* (n.d.), 255–56, American Book Company Archives (ABC Archives), Syracuse University Library, Syracuse, NY. Brammer observes that Barnes called this the School Book Publishers Board of Trade. The words "School Book" appear to have been a conflation with the name of the School Book Publishers Association.

22. Ibid., 260–61.

23. Ibid., 264 (quotations), 267, 268.

24. "American School-Book Publishing," *Publishers' Weekly*, no. 912 (July 20, 1889): 46a.

25. Brammer, *Our Heritage and History*, 276.

26. "The New School-Book Combination," *Publishers' Weekly*, no. 952 (April 26, 1890): 557.

27. George R. Cathcart quoted in "The American Book Company," *Publishers' Weekly*, no. 952 (April 26, 1890): 558.

28. *Publishers' Weekly*, no. 957 (May 31, 1890): 717.

29. Eugene Exman, *The House of Harper* (New York: Harper & Row, 1967), 171–72.

30. John Tebbel, *Between Covers: The Rise and Transformation of Book Publishing in America* (New York: Oxford University Press, 1987), 94.

31. Richard Rogers Bowker, *Copyright: Its History and Its Law* (London: Constable, 1912), 341–72.

32. Brammer, *Our Heritage and History*, 303.

33. "The Olympia Book Bribery Case," *Chicago Tribune*, June 14, 1890, 5.

34. "The Olympia School-Book Bribery," *Chicago Tribune*, June 12, 1890, 2.

35. "Circular, November 24, 1890, by the State Superintendent of Mississippi," quoted in Jeremiah Jenks, "School-Book Legislation," *Political Science Quarterly* 6, no. 1 (March 1891): 100–101.

36. Ginn, *Are Our Schools in Danger?* 2.

37. American Book Co., *Who Is Edwin Ginn?* (New York: American Book Co., 1895), 1–2, 5.

38. George A. Gates, *A Foe of American Schools* (Minneapolis: Kingdom, 1897).

39. *American Book Company v. Kingdom Publishing and Others*, 71 Minn. 363 (1898).

40. *American Book Company Vindicated: The "Gates Pamphlet" a Libel* (n.p., 1898).

41. American Book Company letter, Pamphlets on Education, 1861–96, microfilm, California Department of Public Instruction.

42. Brammer, *Our Heritage and History*, 329, 439 (quotation).

43. Ginn, *Are Our Schools in Danger?* 3.

44. Amanda Meeker, "Overview of the History of Constitutional Provisions Dealing with K-12 Education" (n.d.), Reports of the California Constitutional Revision Commission, http://www.worldcat.org/arcviewer/1/CAX/2006/06/05/0000020402/viewer/file181.html (accessed March 6, 2012).

45. McDougal quoted in J. Ross Browne, *The Debates in the Convention of California, on the Formation of the State Constitution* (Washington, DC: John T. Towers, 1850), 353.

46. Bureau of the Census, *Ninth Census of the United States*, 3 vols. (Washington, DC: U.S. Government Printing Office, 1872), 1:799.

47. Browne, *The Debates in the Convention of California*, 2:1108 (O'Sullivan quotation), 3:1400.

48. Ibid., 3:1401

49. Percy Roland Davis, "State Publication of Textbooks in California" (Ph.D. diss., University of California, Berkeley, 1930).

50. Winfield J. Davis, *History of Political Conventions in California, 1849–1892* (Sacramento: California State Library, 1893); Adam R. Shapiro, "State Regulation of the Textbook Industry," in *Education and the Culture of Print in Modern America*, ed. Adam R. Nelson and John L. Rudolph (Madison: University of Wisconsin Press, 2010), 178–82.

51. Jenks, "School-Book Legislation," 91.

52. "State Text-Book Laws and Systems," *Publishers' Weekly*, no. 1068 (July 16, 1892): 108 (quotation), 109–10. The thirteen states were Indiana, Kentucky, Louisiana, Minnesota, Missouri, Montana, Nevada, North Carolina, Oregon, South Carolina, Virginia, Washington, and West Virginia.

53. *Kansas v. American Book Company*, 69 Kan. 1 (1904); John Franklin Brown, *State Publication of Schoolbooks* (New York: Macmillan, 1931), 24.

54. M. L. Brittain, chair, *Report of the School Book Investigating Committee to the General Assembly of Georgia* (n.p., 1914), 22.

55. P. L. Harned, *Textbook Law of Tennessee* (Nashville: State of Tennessee, 1927), 5.

56. John A. Lapp, "Legislative Notes and Reviews," *American Political Science Review* 10, no. 3 (August 1916): 546.

57. Brammer, *Our Heritage and History*, 439.

58. Ohio School Book Commission, *Report of the State School Book Commission for the Year 1905* (Columbus: State of Ohio, 1905), 1.

59. Bureau of the Census, *Fourteenth Census of the United States*, 11 vols. (Washington, DC: U.S. Government Printing Office, 1922–25), 2:1047.

60. Moreau, *Schoolbook Nation*, 92–136.

61. Brown, *State Publication of Schoolbooks*, 5.

62. Lewis B. Avery, "State-Printed Textbooks in California," *Elementary School Journal* 19, no. 8 (April 1919): 628, 631.

63. *San Francisco Examiner*, December 4, 1890, quoted in Jenks, "School-Book Legislation," 105–6.

64. Avery, "State-Printed Textbooks in California," 628.

65. Davis, "State Publication of Textbooks in California," 38–39.

66. The reference here is to Ellen M. Cyr's *The Children's Primer* (Boston: Ginn, 1891), which was released in California as *The Children's Primer, Compiled by the State Text-Book Committee and Approved by the State Board of Education* (Sacramento, CA: State Printing Office, 1905).

67. Davis, "State Publication of Textbooks in California," 33.

68. *Senate Daily Journal* quoted in Brown, *State Publication of Schoolbooks*, 44–45.

69. W. T. H. Howe to W. W. Livengood, March 19, 1924, ABC Archives.

70. W. W. Livengood to W. T. H. Howe, March 25, 1924, ABC Archives.

71. J.M.S. [Jessie M. Shaver], review of *New Civic Biology*, by George W. Hunter, *Peabody Journal of Education* 7, no. 1 (July 1929): 57.

72. *Report of the Special Senate Investigating Committee*, 163.

73. Ibid.

74. *Nashville Tennessean*, March 17, 1925, quoted in Brown, *State Publication of Schoolbooks*, 27–28, 28.

75. Harned, *Textbook Law of Tennessee*, 5.

76. "Hearing from the Different Book Companies," from the minutes of the state textbook commission (June 1919), folders 2 and 4, box 191, Records Group 92, Commissioner of Education Records, 1913–70, TSLA.

77. Edward J. Larson, *Trial and Error: The American Controversy over Creation and Evolution* (New York: Oxford University Press, 1989), 50–52.

78. Austin M. Peay to Tennessee House, printed in *Journal of Tennessee House, 1925* (Nashville: State of Tennessee, 1925), 743–45.

CHAPTER THREE

1. Charles J. St. John to W. T. H. Howe, December 20, 1923, American Book Company Archives (ABC Archives), Syracuse University Library, Syracuse, NY.

2. Austin M. Peay to Tennessee House, printed in *Journal of Tennessee House, 1925* (Nashville: State of Tennessee, 1925), 744–45.

3. H. L. Donovan, "How to Select Textbooks," *Peabody Journal of Education* 2, no. 1 (July 1924): 2.

4. David R. Olsen, "On the Language and Authority of Textbooks," *Journal of Communication* 30, no. 1 (March 1980): 192.

5. Adrian Johns, *The Nature of the Book: Print and Knowledge in the Making* (Chicago: University of Chicago Press, 1998).

6. Lorraine Daston and Peter Galison, *Objectivity* (New York: Zone, 2007), 309–62.

7. Ibid., 17.

8. For a severe example of this, see the account given in Thomas Woody, "The Country Schoolmaster of Long Ago," *History of Education Journal* 5, no. 2 (Winter 1954): 42–43.

9. Adam R. Shapiro, "Between Training and Popularization: Regulating Science Textbooks in Secondary Education," *Isis* 103, no. 1 (March 2012): 99–110.

10. Charles J. St. John to W. T. H. Howe, December 20, 1923, ABC Archives.

11. In addition to those examples mentioned in the previous chapter, others include *Answer of the American Book Company to the False Charges First Printed in the Norfolk Pilot and Reprinted in the Chicago Inter-Ocean* (n.p., 1895), in which the ABC alleges that copies of a libelous accusation against it were the work of rival publishers, and *The Adoption of the Pacific Coast Series of Readers* (San Francisco: A. L. Bancroft, 1875), in which Bancroft prints a list of testimonials and analyses demonstrating the superiority of its texts for use in California.

12. W. T. H. Howe to George W. Benton, May 2, 1923, ABC Archives.

13. Agents' Meeting, memorandum, December 1926, 1, 2, George A. Plimpton Papers, Rare Book and Manuscript Library, Columbia University, New York.

14. George A. Plimpton, address to firm meeting, January 1926, 9–10, Plimpton Papers.

15. Walter A. Friedman, *Birth of a Salesman: The Transformation of Selling in America* (Cambridge, MA: Harvard University Press, 2004), 10 (quotation), 37–49.

16. Thomas Herbert Russell, *Salesmanship Theory and Practice* (Toronto: International Law and Business Institute, 1910), 15.

17. Edward K. Strong Jr., *The Psychology of Selling and Advertising* (New York: McGraw-Hill, 1925), 8.

18. Russell, *Salesmanship Theory and Practice*, 15.

19. William Jessup Sholar, ed., *Salesmanship: The Standard Course of the United Y.M.C.A. Schools* (New York: Association Press, 1920), 3.

20. Thomas Russell, *An Introduction to Sales Management* (London: Duckworth, 1924), 12.

21. Friedman, *Birth of a Salesman*, 118.

22. Peter Filene, "The World Peace Foundation and Progressivism: 1910–1918," *New England Quarterly* 36, no. 4 (December 1963): 483.

23. "Address of Mr. Edwin Ginn," in *Report of the Seventh Annual Meeting of the Lake Mohonk Conference on International Arbitration* (Albany, NY: Lake Mohonk Arbitration Conference, 1901), 22.

24. Arthur N. Holcombe, "Edwin Ginn's Vision of World Peace," *International Organization* 19, no. 1 (Winter 1965): 2.

25. Ibid., 1–2.

26. John White, "Andrew Carnegie and Herbert Spencer: A Special Relationship," *Journal of American Studies* 13, no. 1 (April 1979): 57–71; Andrew Carnegie, *Autobiography of Andrew Carnegie, with illustrations* (Boston: Houghton Mifflin, 1920), 333.

27. William Jennings Bryan, *The Forces That Make for Peace: Addresses at the Mohonk Conferences on International Arbitration, 1910 and 1911* (Boston: World Peace Foundation, 1912).

28. Edward J. Larson, *Summer for the Gods: The Scopes Trial and America's Continuing Debate over Science and Religion* (Cambridge, MA: Harvard University Press, 1997), 40; Lawrence W. Levine, *Defender of the Faith: William Jennings Bryan: The Last Decade, 1915–1925* (New York: Oxford University Press, 1965), 262.

29. F. M. Ambrose to the Managing Committee, December 18, 1912, Plimpton Papers.

30. Thomas Bonaventure Lawler, *Seventy Years of Textbook Publishing* (Boston: Ginn, 1938), 19.

31. Richard S. Thomas to the Agents, July 15, 1915, Plimpton Papers.

32. Richard S. Thomas to the Offices, July 15, 1915, Plimpton Papers.

33. "The Iroquois Publishing Company, Inc.," advertising poster, 1939, Collections of the Onondaga Historical Association, Syracuse, NY.

34. Sarah Louise Arnold, Elizabeth C. Bonney, and E. F. Southworth, *The See and Say Series* (Boston: Ginn, 1913).

35. M. C. Collister and O. J. Walrath, *Laboratory Guide for Chemistry* (Syracuse, NY: Iroquois, 1922); Roswell B. Peters, *Laboratory Guide for Biology* (Syracuse, NY: Iroquois, 1922).

36. "Apportionment of Agents' Time as to States, New York Field," memorandum, November 1915, Plimpton Papers.

37. Arthur G. Clement, *Living Things: An Elementary Biology* (Syracuse, NY: Iroquois, 1924).

38. E. F. Southworth to Austin Peay, n.d., folder 1, container 44, Governor Austin Peay Papers, Tennessee State Library and Archives, Nashville. Southworth's letter was written on the stationary, not of Iroquois Publishing, but of the Hotel Hermitage in Nashville. See also George W. Hunter, *A Civic Biology, Presented in Problems* (New York: American Book Co., 1914); and William M. Smallwood, Ida L. Reveley, and Guy A. Bailey, *Practical Biology* (Boston: Allyn & Bacon, 1916).

39. E. F. Southworth to "The Head of the Biology Department and the Teachers of Biology," May 28, 1924, container 13, Benjamin C. and Sidonie M. Gruenberg Papers, Manuscripts Division, Library of Congress, Washington, DC.

40. Henry R. Linville to Benjamin C. Gruenberg, June 3, 1924, container 13, Gruenberg Papers. See also Philip J. Pauly, "The Development of High School Biology: New York City, 1900–1925," *Isis* 82, no. 4. (December 1991): 662–88.

41. Avery W. Skinner to Benjamin C. Gruenberg, April 23, 1923, container 13, Gruenberg Papers.

42. William M. Smallwood to Benjamin C. Gruenberg, June 26, 1924, container 13, Gruenberg Papers.

43. George W. Hunter to Benjamin C. Gruenberg, September 19, 1924, container 13, Gruenberg Papers.

44. James E. Peabody to Benjamin C. Gruenberg, June 19, 1924, container 13, Gruenberg Papers.

45. Benjamin C. Gruenberg to O. W. Caldwell, March 24, 1923, container 13, Gruenberg Papers.

46. Benjamin C. Gruenberg to Ida L. Reveley et al., June 9, 1924, container 13, Gruenberg Papers.

47. James E. Peabody to Benjamin C. Gruenberg, August 19, 1924, container 13, Gruenberg Papers.

48. James E. Peabody to Benjamin C. Gruenberg, January 30, 1912, container 95, Gruenberg Papers.

49. Frank M. Wheat and Elizabeth T. Fitzpatrick, *Advanced Biology* (New York: American Book Co., 1929), iv.

50. Elizabeth T. Fitzpatrick to Frank M. Wheat, January 23, 1922, container 13, Gruenberg Papers.

51. Pauly, "The Development of High School Biology," 669.

52. W. T. H. Howe to George W. Benton, May 2, 1923, ABC Archives.

53. Ibid.

54. L. B. Lee to W. W. Livengood, June 14, 1924, ABC Archives.

55. For examples of Hunter's interactions with ABC employees, see chapter 6.

56. Johns, *The Nature of the Book*, 3.

57. Charles J. St. John to W. T. H. Howe, December 20, 1923, ABC Archives.

CHAPTER FOUR

1. Frank A. Fitzpatrick to George W. Benton, February 19, 1915, American Book Company Archives (ABC Archives), Syracuse University Library, Syracuse, NY.

2. Walter G. Whitman to George W. Hunter, March 15, 1915, ABC Archives.

3. George W. Hunter to Walter G. Whitman, March 16, 1915, ABC Archives.

4. Ada L. McKel, "Suggested Changes in Text of Hunter's Civic Biology to Meet Boston Criticism," memorandum to George W. Hunter, July 3, 1915, ABC Archives.

5. George W. Benton to J. R. Fairchild, July 19, 1915, ABC Archives.

6. W. W. Livengood to George W. Benton, September 17, 1915, ABC Archives.

7. J. R. Fairchild to George W. Benton, April 6, 1915, ABC Archives.

8. George F. Atkinson to W. W. Drew, September 30, 1915, ABC Archives.

9. Elliot Downing, "A Review of the Year's Progress in High-School Science," *School Review* 27, no. 6 (June 1919): 484.

10. Stephen Jay Gould, "An Essay on a Pig Roast," in *Bully for Brontosaurus* (New York: Norton, 1991), 428–29; Edward J. Larson, *Summer for the Gods: The Scopes Trial and America's Continuing Debate over Science and Religion* (Cambridge, MA: Harvard University Press, 1997), 27–28.

11. I use the term *Fundamentalism* to refer specifically to the movement spurred by *The Fundamentals* during the 1910s and 1920s and the term *fundamentalism* to refer to religious extremism in general. *The Fundamentals* was privately published in twelve paperback volumes from 1910 to 1915, edited first by A. C. Dixon and later by R. A. Torrey.

12. Ronald L. Numbers, *Darwinism Comes to America* (Cambridge, MA: Harvard University Press, 1998).

13. William M. Smallwood, Ida L. Reveley, and Guy A. Bailey, *Practical Biology* (Boston: Allyn & Bacon, 1916); P. L. Harned, *List and Prices of Text Books Adopted in 1919 and Prices on the Same Books from September 1, 1924 to June 30, 1925* (Nashville: State of Tennessee, 1924), 7.

14. In 1924–25, 3,274 students used Hunter's book in Tennessee, compared to 299 students who used the *Practical Biology* (figures based on data compiled from Principal's Annual Reports, 1924–25, boxes 66–67, Department of Education Records, 1874–1964, Record Group 273, Tennessee State Library and Archives [TSLA], Nashville). Only school annual reports that listed both a textbook and the number of students taking biology are included in these counts. It does not appear that segregated schools for black students in Tennessee taught biology at this time.

15. The first textbook published in the United States to use the word *biology* in its title was T. H. Huxley and H. N. Martin's *Course of Practical Instruction in Elementary Biology* (New York: Macmillan, 1876).

16. William T. Sedgwick and Edmund B. Wilson, *General Biology* (New York: Henry Holt, 1886), iii–iv.

17. Philip J. Pauly, "The Development of High School Biology: New York City, 1900–1925," *Isis* 82, no. 4 (December 1991): 662–88.

18. Ibid., 663, 667.

19. *Catalogue of Williams College, 1892–1893*, 35, Williams College Archives, Williamstown, MA. The course catalog describes the introductory biology course as "Sophomore required course" and "an introductory study consisting of lectures supplemented by lessons from Sedgwick and Wilson's 'Biology,' and by laboratory practice." Hunter was a sophomore in 1892–93.

20. University of Chicago Transcript of George W. Hunter, University of Chicago Archives, Chicago.

21. *The Clintonian*, yearbook of DeWitt Clinton High School (New York, 1913), 76.

22. Adam R. Shapiro, "Civic Biology and the Origin of the School Antievolution Movement," *Journal of the History of Biology* 41, no. 3 (2008): 418.

23. Felix Adler, *The Moral Instruction of Children* (New York: D. Appleton, 1892), 3.

24. *Catalogue of Officers and Graduates of Columbia University* (New York: Columbia University, 1906), 573; Stephen T. Correia, "A Small Circle of Friends: Clarence Kingsley and Membership on the Committee of the Commission on the Reorganization of Secondary Education," ERIC Document Reproduction Service No. ED398136 (1995), 5.

25. Benjamin C. Gruenberg, "Some Aspects of the Child-Welfare Problem in the New York High Schools," *School Review* 19, no. 10 (December 1911): 684.

26. Philip J. Pauly, *Biologists and the Promise of American Life: From Meriwether Lewis to Alfred Kinsey* (Princeton, NJ: Princeton University Press, 2002), 178n; Otis W. Caldwell and William L. Eikenberry, *Elements of General Science* (Boston: Ginn, 1914).

27. William G. Wraga, "Who Wrote the Cardinal Principles Report? The Commission on the Reorganization of Secondary Education Revisited," ERIC Document Reproduction Service No. ED454603 (1999), 3.

28. Shapiro, "Civic Biology and the Origin of the School Antievolution Movement," 418–19.

29. Larson, *Summer for the Gods*, 73–81.

30. Benjamin C. Gruenberg to Commissioner of Education, November 30, 1921, box 13, Benjamin C. and Sidonie M. Gruenberg Papers, Manuscripts Division, Library of Congress, Washington, DC.

31. "Unpatriotic Teaching in the Public Schools," in *New York Legislative Documents*, vol. 19, no. 50, pt. 3 (Albany, NY: J. B. Lyon, 1921), 2665–67.

32. Henry Richardson Linville Papers, Walter P. Reuther Library, Wayne State University, Detroit, MI; Clarence M. Pruitt, "Benjamin Charles Gruenberg," *Science Education* 50, no. 1 (February 1966): 83–89.

33. *Reorganization of Science in Secondary Schools*, Department of the Interior, Bureau of Education, Bulletin, 1920, no. 26 (Washington, DC: U.S. Government Printing Office, 1920), 13–14, 14.

34. John M. Heffron, "Knowledge Most Worth Having: Otis W. Caldwell (1869–1947) and the Rise of the General Science Course," *Science and Education* 4 (1995): 235.

35. *Reorganization of Science in Secondary Schools*, 15.

36. John L. Rudolph, "Turning Science to Account: Chicago and the General Science Movement in Secondary Education, 1905–1920," *Isis* 96, no. 3 (2005): 354–55.

37. George W. Hunter, *A Civic Biology, Presented in Problems* (New York: American Book Co., 1914), 373–97.

38. Herbert Spencer, *Education: Intellectual, Moral, Physical* (1860; reprint, New York: D. Appleton, 1898), 123.

39. Adler, *Moral Instruction of Children*, 251, 251–53.

40. John Dewey, *Democracy and Education* (1916; reprint, New York: Macmillan, 1922), 85.

41. Heffron, "Knowledge Most Worth Having," 227; Andrew Feffer, *Chicago Pragmatists and American Progressivism* (New York: Cornell University Press, 1993), 150n.

42. Robert A. Levin, "The Debate over Schooling: Influences of Dewey and Thorndike," *Childhood Education* 6, no. 2 (Winter 1991): 71–75.

43. Edward L. Thorndike, *Educational Psychology*, vol. 1, *The Original Nature of Man* (New York: Teacher's College, 1913), vii.

44. G. E. Partridge, *Genetic Philosophy of Education: An Epitome of the Published Writings of President G. Stanley Hall of Clark University* (New York: Sturgis & Walton, 1912), 256, 255–56.

45. Arthur N. Holcombe, "Edwin Ginn's Vision of World Peace," *International Organization* 19, no. 1 (Winter 1965): 1–2.

46. George W. Hunter, *Essentials of Biology* (New York: American Book Co., 1911), 418–36.

47. Hunter, *Civic Biology*, 9.

48. Pauly, "The Development of High School Biology," 664.

49. Shapiro, "Civic Biology and the Origin of the School Antievolution Movement," 419.

50. Hunter, *Civic Biology*, 9.

51. Pauly, "The Development of High School Biology," 662–88.

52. William E. Cole, *The Teaching of Biology* (New York: D. Appleton, 1934), 12.

53. Pauly, "The Development of High School Biology," 664.

54. Walter G. Whitman, report sent to George W. Benton, n.d. [1914], ABC Archives.

55. George W. Hunter, *New Essentials of Biology* (New York: American Book Co., 1923), vi.

56. Rudolph, "Turning Science to Account."

57. George W. Hunter, "The Sequence of Science in the Junior and Senior High School," *High School Journal* 3, no. 6 (October 1920): 163–65, and "The Place of Science in the Secondary School, I," *School Review* 33, no. 5 (May 1925): 370–81.

58. Keith Sheppard and Dennis M. Robbins, "A History of the Grade Placement of High School Biology," *American Biology Teacher* 68, no. 7 (2006): 86–90.

59. Hunter, *New Essentials of Biology*, v.

60. Rudolph, "Turning Science to Account," 356.

61. George W. Benton to George W. Hunter, October 31, 1924, ABC Archives.

62. Charles A. Israel, *Before Scopes: Evangelism, Education, and Evolution in Tennessee, 1870–1925* (Athens: University of Georgia Press, 2004), 97–127.

63. James H. Blodgett, "The Rural School Problem," *American Anthropologist* 6, no. 1 (January 1893): 71–78.

64. Tracy Steffes, "Solving the 'Rural School Problem': New State Aid, Standards, and Supervision of Local Schools, 1900–1933," *History of Education Quarterly* 48, no. 2 (May 2008): 192.

65. P. L. Harned to Members of Senate, memorandum, n.d., folder 5, container 98, Governor Austin Peay Papers, TSLA.

66. Jeanette Keith, *Country People in the New South* (Chapel Hill: University of North Carolina Press, 1995), 202, 203.

67. Austin M. Peay to Tennessee House, printed in *Journal of the Tennessee House, 1925* (Nashville: State of Tennessee, 1925), 744, 744–45.

68. Kenneth K. Bailey, "The Enactment of Tennessee's Antievolution Law," *Journal of Southern History* 16, no. 4 (November 1950): 478–82.

69. Larson, *Summer for the Gods*, 57.

70. P. P. Claxton to Austin M. Peay, March 10, 1925, folder 7, container 17, Peay Papers.

71. Paul K. Conkin, *When All the Gods Trembled: Darwinism, Scopes, and American Intellectuals* (Lanham, MD: Rowman & Littlefield, 1998), 82.

72. *Journal of the Tennessee House, 1925*, 375, 387, 655, 804, 806–7, 1049.

73. *Journal of the Tennessee Senate, 1925* (Nashville: State of Tennessee, 1925), 1263, 1290.

74. Mrs. C. B. Allen to Austin M. Peay, April 13, 1925, folder 3, container 59, Peay Papers.

75. "Peay Appoints Woman," *Memphis Commercial Appeal*, April 21, 1925, 6.

76. Mrs. C. B. Allen to P. L. Harned, March 2, 1925, folder 3, container 59, Peay Papers.

77. Mrs. C. B. Allen to P. L. Harned, April 16, 1925, folder 3, container 59, Peay Papers.

78. "Peay Appoints Woman."

79. William Jennings Bryan, "God and Evolution," *New York Times*, February 26, 1922, 1, 11.

80. Bailey, "The Enactment of Tennessee's Antievolution Law."

81. Clarence Darrow to Benjamin C. Gruenberg, telegram, July 10, 1925, container 17, Gruenberg Papers.

82. Marcel Chotkowski LaFollette, *Reframing Scopes: Journalists, Scientists and Lost Photographs from the Trial of the Century* (Lawrence: University Press of Kansas, 2008), 43n.

83. Benjamin C. Gruenberg to E. N. Stevens, July 10, 1925, container 17, Gruenberg Papers.

84. Benjamin C. Gruenberg to Clarence Darrow, telegram, July 10, 1925, container 17, Gruenberg Papers.

85. L. D. Roberson to Benjamin C. Gruenberg, telegram, July 11, 1925, container 17, Gruenberg Papers.

86. Benjamin C. Gruenberg to Clarence Darrow, telegram, July 13, 1925, container 17, Gruenberg Papers.

87. Pruitt, "Benjamin Charles Gruenberg," 85.

88. F. C. Hodgson to Benjamin C. Gruenberg, June 22, 1925, container 17, Gruenberg Papers.

89. Joel N. Pierce quoted in "Says School Board Avoided Argument," *Chattanooga Times*, June 14, 1925, 3.

90. John R. Neal to Austin Peay, June 1, 1925, folder 2, container 44, Peay Papers.

91. Austin Peay to P. L. Harned, n.d., Peay Papers, folder 2, container 44, Peay Papers.

92. "Co-Author Amazed at His Book's Part," *Chattanooga Times*, June 15, 1925, 5.

93. David B. Wilson, "The Historiography of Science and Religion," in *Science and Religion: A Historical Introduction*, ed. Gary B. Ferngren (Baltimore: Johns Hopkins University Press, 2002), 13–29; Geoffrey Cantor, "What Shall We Do with the 'Conflict Thesis?'" in *Science and Religion: New Historical Perspectives*, ed. Thomas Dixon, Geoffrey Cantor, and Stephen Pumfrey (Cambridge: Cambridge University Press, 2010), 283–98.

CHAPTER FIVE

1. William Jennings Bryan to C. H. Thurber, December 22, 1923, container 38, William Jennings Bryan Papers, Manuscripts Division, Library of Congress, Washington, DC.

2. William Jennings Bryan, "God and Evolution," *New York Times*, February 26, 1922, 1.

3. Ibid.

4. James H. Leuba, *The Belief in God and Immortality* (Boston: Sherman, French, 1916).

5. Jon Roberts, "Science and Religion," in *Wrestling with Nature: From Omens to Science*, ed. Peter Harrison, Ronald L. Numbers, and Michael H. Shank (Chicago: University of Chicago Press, 2011), 254.

6. *Journal of Tennessee House, 1925* (Nashville: State of Tennessee, 1925), 1455.

7. Jeanette Keith, *Country People in the New South* (Chapel Hill: University of North Carolina Press, 1995), 27.

8. Charles A. Israel, *Before Scopes: Evangelicals, Education, and Evolution in Tennessee, 1870–1925* (Athens: University of Georgia Press, 2004), 11–42.

9. John Washington Butler, "A Brief History of the Anti-Evolution Law of Tennessee," April 2, 1927, in microfilm 485, John Washington Butler, Biographical Questionnaires, ca. 1800–1922, Tennessee State Library and Archives, Nashville.

10. *Scopes v. State*, 154 Tenn. 105 (1927).

11. John Washington Butler quoted in "Author of the Law Surprised at the Fuss," *New York Times*, July 18, 1925, 1.

12. Keith, *Country People in the New South*, 200–201.

13. "Author of the Law Surprised at the Fuss," 2.

14. Austin M. Peay to Tennessee House, printed in *Journal of Tennessee House, 1925*, 744–45.

15. Paul K. Conkin, *When All the Gods Trembled: Darwinism, Scopes, and American Intellectuals* (Lanham, MD: Rowman & Littlefield, 1998), 82–83.

16. *Public Acts of the State of Tennessee, Passed by the Sixty-fourth General Assembly, 1925* (Nashville: Tennessee Industrial School, 1925), 50–51. Sections 2 and 3 concern the penalties for violating and the date of implementing the law.

17. E. B. Ewing, stenog., "State of Tennessee v. J. T. Scopes," in *The Scopes Case* (Wilmington, DE: Michael Glazier, 1978), microfilm, 280.

18. Edward J. Larson, *Trial and Error: The American Controversy over Creation and Evolution* (New York: Oxford University Press, 1989), 49–52.

19. John T. Scopes and James Presley, *Center of the Storm: Memoirs of John T. Scopes* (New York: Holt, Rinehart & Winston, 1967), 58–59.

20. Edward J. Larson, *Summer for the Gods: The Scopes Trial and America's Continuing Debate over Science and Religion* (Cambridge, MA: Harvard University Press, 1997), 65–73.

21. Clarence Darrow, *The Story of My Life* (1932; reprint, New York: Da Capo, 1996), 244.

22. L. D. Roberson to Benjamin C. Gruenberg, July 11, 1925, container 17, Benjamin C. and Sidonie M. Gruenberg Papers, Manuscripts Division, Library of Congress, Washington, DC; George W. Benton to George W. Hunter, June 10, 1925, American Book Company Archives (ABC Archives), Syracuse University Library, Syracuse, NY.

23. Larson, *Summer for the Gods*, 139–44.

24. "Cranks and Freaks Flock to Dayton," *New York Times*, July 11, 1925, 1.

25. Michael Lienesch, *In the Beginning: Fundamentalism, the Scopes Trial, and the Making of the Antievolution Movement* (Chapel Hill: University of North Carolina Press, 2007), 141.

26. "Dayton Keyed Up for Opening Today of Trial of Scopes," *New York Times*, July 10, 1925, 1.

27. Mary Beth Swetnam Mathews, *Rethinking Zion: How the Print Media Placed Fundamentalism in the South* (Knoxville: University of Tennessee Press, 2006), 84.

28. Constance Clark, *God—or Gorilla: Images of Evolution in the Jazz Age* (Baltimore: Johns Hopkins University Press, 2008), 168, 132 (see generally 132–61).

29. Susan Harding, "Representing Fundamentalism: The Problem of the Repugnant Cultural Other," *Social Research* 58, no. 2 (Summer 1991): 377, 374.

30. The depiction of the Scopes trial in the Creation Museum in Petersburg, Kentucky, exemplifies this binary opposition.

31. Ewing, "State of Tennessee v. J. T. Scopes," 301 (quotation), 311–23.

32. F. E. Robinson and W. E. Morgan, *Dayton's Cultural Growth, Particularly Agri-Cultural!* (Chattanooga: Andrews Printery, 1925).

33. Ewing, "State of Tennessee v. J. T. Scopes," 324.

34. Ibid., 537–47.

35. Ibid., 288.

36. *Epperson v. Arkansas*, 393 U.S. 97 (1968); *McLean v. Arkansas Board of Education*, 529 F. Supp. 1255 (1982); *Kitzmiller et al. v. Dover Area School District*, 400 F. Supp. 2d 707 (2005).

37. *Everson v. Board of Education*, 330 U.S. 1 (1947).

38. *Scopes v. State*, 154 Tenn. 105 (1927). The majority opinion rejected this defense argument.

39. *Connally v. General Const. Co.*, 269 U.S. 385 (1926).

40. *Theistic evolution* has been criticized as an imprecise description of several religious positions that do not reject evolution and that accept the existence of a god. Despite its imprecisions, it is a term that participants on both sides of the Scopes trial used, and it is therefore used here. See Antje Jackelén, "A Critical View of 'Theistic Evolution,'" *Theology and Science* 5, no. 2 (July 2007): 151–65; and Ted Peters and Martinez Hewlett, *Can You Believe in God and Evolution?* (Nashville: Abingdon, 2006).

41. In their appeal to the Tennessee state supreme court, Scopes's lawyers also argued that the antievolution law violated the provision of the state constitution stating that the legislature had a duty "to cherish literature and science." The court found this provision too vague to apply to the law. This provision of the state constitution has since been revoked.

42. Shailer Matthews quoted in "Science and Fundamentalism," *Chicago Tribune*, April 1, 1925, 8.

43. "Dr. J. Frank Norris to Reach Here Today," *Memphis Commercial Appeal*, May 2, 1925, 14.

44. Larson, *Summer for the Gods*, 99.

45. Ibid., 81–83.

46. William Jennings Bryan quoted in "Bryan to Help in Legal Fight over Evolution," *Chicago Tribune*, May 13, 1925, 3.

47. Larson, *Summer for the Gods*, 44–45.

48. Lawrence W. Levine, *Defender of the Faith: William Jennings Bryan: The Last Decade, 1915–1925* (New York: Oxford University Press, 1965), 360.

49. John Thomas Scopes, "Reflections—*Forty Years After*," in *D-Days at Dayton*, ed. Jerry R. Tompkins (Baton Rouge: Louisiana State University Press, 1965), 19–20.

50. Ewing, "State of Tennessee v. J. T. Scopes," 280–81.

51. Ibid., 346.

52. Ibid., 371.

53. Scopes, "Reflections—*Forty Years After*," 25–26.

54. Ewing, "State of Tennessee v. J. T. Scopes," 444.

55. William Jennings Bryan to W. B. Marrs [Marr], June 11, 1925, copy, enclosed in Ewing C. Baskette to Clarence Darrow, January 6, 1934, box 2, Clarence Darrow Papers, Manuscripts Division, Library of Congress, Washington, DC.

56. Ewing, "State of Tennessee v. J. T. Scopes," 451, 452.

57. Ibid., 785.

58. William Jennings Bryan, *In His Image* (New York: Fleming H. Revell, 1922), 127.

59. Ewing, "State of Tennessee v. J. T. Scopes," 744.

60. Ibid., 473.

61. Ibid., 725.

62. Ibid., 727.

63. Ibid., 733.

64. Ibid., 734.

65. Ibid., 789.

66. See Larson, *Summer for the Gods*, 3–8; Michael Kazin, *A Godly Hero: The Life of William Jennings Bryan* (New York: Knopf, 2006), 290–95; Levine, *William Jennings Bryan: The Last Decade*, 348–51.

67. Ewing, "State of Tennessee v. J. T. Scopes," 801.

68. Ronald L. Numbers, *The Creationists*, expanded ed. (Cambridge, MA: Harvard University Press, 2003), 51–87.

69. Austin M. Peay to Tennessee House, printed in *Journal of Tennessee House, 1925*, 744–45.

70. Scopes and Presley, *Center of the Storm*, 59.

71. George W. Hunter, *A Civic Biology, Presented in Problems* (New York: American Book Co., 1914), 195.

72. Benjamin C. Gruenberg, *Elementary Biology* (Boston: Ginn, 1919), 494–97.

73. Constance Clark, "Evolution for John Doe: Pictures, the Public, and the Scopes Trial Debate," *Journal of American History* 87, no. 4 (March 2001): 1275–77.

74. Ewing, "State of Tennessee v. J. T. Scopes," 328.

75. Clark, "Evolution for John Doe," 1276–77.

76. Clark, *God—or Gorilla*, 132–94, and "Evolution for John Doe," 1279.

77. Hunter, *Civic Biology*, 194–95.

CHAPTER SIX

1. George W. Hunter to George W. Benton, May 25, 1925, and Hunter to Benton, June 2, 1925, American Book Company Archives (ABC Archives), Syracuse University Library, Syracuse, NY.

2. "Teacher Tests Evolution Law in Tennessee," *Chicago Tribune*, May 7, 1925, 12. All other examples are from newspaper clippings dated May 7, 1925, in ABC Archives.

3. "Held Trial for Teaching Evolution," *Chicago Tribune*, May 10, 1925, 3.

4. George W. Benton to Louis B. Lee, April 23, 1925, ABC Archives.

5. Harry D. Nutt to Louis B. Lee, June 7, 1925, ABC Archives.

6. George W. Benton to George W. Hunter, June 8, 1925, ABC Archives.

7. George W. Benton to George W. Hunter, June 10, 1925, ABC Archives.

8. Judith V. Grabiner and Peter D. Miller, "Effects of the Scopes Trial: Was It a Victory for Evolutionists?" *Science* 185 (1974): 833.

9. Randy Moore, "The Lingering Impact of the Scopes Trial on High School Biology Textbooks," *BioScience* 51, no. 9 (September 2001): 790–96.

10. "Says School Board Avoided Argument," *Chattanooga Times*, June 14, 1925, 3.

11. Grabiner and Miller, "Effects of the Scopes Trial," 833.

12. Edward J. Larson, *Summer for the Gods: The Scopes Trial and America's Continuing Debate over Science and Religion* (Cambridge, MA: Harvard University Press, 1997), 230–31.

13. Henry William Elson, *Modern Times and the Living Past* (New York: American Book Co., 1921), 2. The relevant quotation reads: "Man is the only creature that has a moral and religious instinct."

14. N. B. Rose to W. T. H. Howe, April 7, 1926, ABC Archives.

15. Charles H. Thurber to William Jennings Bryan, November 21, 1923, container 38, William Jennings Bryan Papers, Manuscripts Division, Library of Congress, Washington, DC.

16. George W. Benton to Frank H. Ellis, April 9, 1915, ABC Archives.

17. W. W. Livengood to Louis B. Lee, August 6, 1920, ABC Archives.

18. *Who's Who in America* 13 (1924–25): 1679, quoted in George W. Benton to W. T. H. Howe, August 6, 1926, ABC Archives.

19. Jarvis R. Fairchild to George W. Benton, August 27, 1920, ABC Archives.

20. B. O. M. DeBeck to W. T. H. Howe, August 30, 1920, ABC Archives.

21. W. W. Livengood to Jarvis R. Fairchild, January 4, 1922, ABC Archives.

22. W. W. Livengood to George W. Hunter, January 28, 1922, ABC Archives.

23. George W. Hunter to W. W. Livengood, February 3, 1922, ABC Archives. Across the top of this letter, Livengood scribbled: "No answer to such a letter as this."

24. George W. Hunter to George W. Benton, February 11, 1924, ABC Archives. Benton wrote of Assistant Editor Stiles A. Torrance's reaction on this letter.

25. George W. Benton, memorandum, June 9, 1925, ABC Archives.

26. George W. Hunter and Walter C. Whitman, *Civic Science in the Home and Community* (New York: American Book Co., 1921).

27. George W. Benton, memorandum, September 6, 1924, ABC Archives.

28. George W. Benton to George W. Hunter, October 31, 1924, ABC Archives.

29. George W. Hunter to George W. Benton, November 3, 1924, ABC Archives.

30. George W. Benton to George W. Hunter, December 9, 1924, ABC Archives.

31. George W. Hunter to George W. Benton, April 22, 1925, ABC Archives.

32. George W. Hunter to George W. Benton, May 15, 1925, ABC Archives.

33. Benjamin C. Gruenberg, *Elementary Biology* (1919; rev., New York: Ginn, 1924).

34. Louis B. Lee to George W. Benton, June 15, 1925, ABC Archives.

35. George W. Benton to Stiles A. Torrance, n.d., ABC Archives.

36. George W. Benton to George W. Hunter, June 23, 1925, ABC Archives.

37. George W. Benton to F. A. Blake, May 27, 1925, ABC Archives.

38. W. W. Livengood to Louis B. Lee, July 6, 1925, ABC Archives.

39. George W. Hunter to Louis B. Lee, October 26, 1925, ABC Archives.

40. George W. Benton to George W. Hunter, May 12, 1925, ABC Archives.

41. James E. Peabody and Arthur E. Hunt, *Biology and Human Life* (New York: Macmillan, 1925), vi.

42. George W. Hunter to George W. Benton, June 15, 1925, ABC Archives. Hunter originally wrote "man condescended from a monkey," crossing out the *con*, an honest, yet perhaps telling, error.

43. George W. Benton to George W. Hunter, June 23, 1925, ABC Archives.

44. George W. Hunter to George W. Benton, June 26, 1925, ABC Archives.

45. George W. Hunter, *A Civic Biology, Presented in Problems* (New York: American Book Co., 1914), 195.

46. Mary E. Robb, "Report on Manuscript for Revision of Hunter's Civic Biology," sent to ABC October 16, 1925, ABC Archives. Robb attributes this to p. 312 of the revised manuscript she reviewed.

47. Constance Clark, *God—or Gorilla: Images of Evolution in the Jazz Age* (Baltimore: Johns Hopkins University Press, 2008), 1–16.

48. George W. Hunter to George W. Benton, June 26, 1925, ABC Archives.

49. George W. Hunter to George W. Benton, July 16, 1925, ABC Archives.

50. Louis B. Lee to Louis Dillman, August 19, 1925, ABC Archives.

51. George W. Hunter to Louis B. Lee, August 10, 1925, ABC Archives.

52. W. W. Livengood to J. J. Bliss, August 20, 1925, ABC Archives.

53. George W. Benton to George W. Hunter, July 23, 1925, ABC Archives.

54. George W. Benton to Louis B. Lee, February 18, 1926, ABC Archives.

55. George W. Benton to George W. Hunter, June 23, 1925, ABC Archives.

56. George W. Hunter to George W. Benton, June 15, 1926, ABC Archives.

57. Annah P. Hazen, "Report on Manuscript for Revision of Hunter's Civic Biology," sent to ABC August 3, 1925, ABC Archives.

58. George W. Benton to George W. Hunter, September 12, 1925, ABC Archives.

59. George W. Hunter to Louis B. Lee, October 6, 1925, ABC Archives.

60. George W. Benton to George W. Hunter, July 23, 1925, ABC Archives.

61. D. Dykes, memorandum, September 15, 1925, ABC Archives.

62. George W. Hunter to George W. Benton, September 16, 1925, ABC Archives.

63. McHenry Rhoads to B. O. M. DeBeck, September 17, 1925, ABC Archives.

64. Hunter, *Civic Biology*, 196.

65. McHenry Rhoads to B. O. M. DeBeck, September 17, 1925, ABC Archives.

66. Louis B. Lee to George W. Benton, October 7, 1925, ABC Archives.

67. Robb, "Report on Manuscript for Revision of Hunter's Civic Biology."

68. George W. Benton to Jarvis R. Fairchild, October 29, 1925, ABC Archives.

69. George W. Benton to George W. Hunter, November 5, 1925, ABC Archives.

70. George W. Hunter to George W. Benton, November 10, 1925, ABC Archives.

71. George W. Benton to George W. Hunter, November 13, 1925, ABC Archives.

72. George W. Hunter to George W. Benton, November 16, 1925, ABC Archives.

73. George W. Hunter to George W. Benton, November 24, 1925, and Benton to Hunter, November 25, 1925, ABC Archives.

74. George F. Washburn to Austin Peay, March 11, 1926, Governor Austin Peay Papers, Tennessee State Library and Archive, Nashville, Tennessee.

75. George W. Hunter to George W. Benton, November 24, 1925, ABC Archives.

76. Jarvis R. Fairchild to George W. Benton, January 7, 1926, ABC Archives.

77. George W. Hunter to George W. Benton, December 3, 1925, ABC Archives.

78. George W. Benton to Louis Dillman, December 7, 1925, ABC Archives.

79. Jarvis R. Fairchild to George W. Benton, December 23, 1925, ABC Archives.

80. George W. Benton to George W. Hunter, December 11, 1925, ABC Archives.

81. George W. Hunter to Stiles A. Torrance, January 26, 1926, ABC Archives.
82. George W. Hunter to Stiles A. Torrance, February 3, 1926, ABC Archives.
83. Louis B. Lee to George W. Benton, October 7, 1925, ABC Archives.
84. W. W. Livengood to George W. Hunter, March 29, 1922, ABC Archives.
85. George W. Hunter to W. T. H. Howe, May 21, 1938, ABC Archives.
86. W. T. H. Howe to Louis Dillman, February 5, 1930, ABC Archives.
87. Jesse M. Shaver, "Report on Hunter, A New Civic Biology (pages 107–223)," mailed to ABC January 15, 1926, ABC Archives.
88. Jesse M. Shaver to B. O. M. DeBeck, January 15, 1926, ABC Archives.
89. Jesse M. Shaver, "Report on Hunter, A New Civic Biology," mailed to ABC February 4, 1926, ABC Archives.
90. J.M.S. [Jesse M. Shaver], review of *New Civic Biology*, by George W. Hunter, *Peabody Journal of Education* 7, no. 1 (July 1929): 57.
91. George W. Benton to W. W. Livengood, January 19, 1926, ABC Archives.
92. George W. Benton to George W. Hunter, January 28, 1926, ABC Archives.
93. George W. Benton to W. T. H. Howe, February 15, 1926, ABC Archives.
94. George W. Benton to W. W. Livengood, handwritten note on copy of Benton to W. T. H. Howe, February 15, 1926, ABC Archives.
95. George W. Hunter to Stiles A. Torrance, February 4, 1926, ABC Archives.
96. W. T. H. Howe to George W. Benton, February 17, 1926, ABC Archives.
97. George W. Hunter to George W. Benton, March 15, 1926, ABC Archives.
98. B. O. M. DeBeck to W. T. H. Howe, March 15, 1926, ABC Archives.
99. George W. Hunter to George W. Benton, March 15, 1926, ABC Archives.
100. George W. Benton to George W. Hunter, March 19, 1926, ABC Archives.
101. George W. Hunter to George W. Benton, March 19, 1926, ABC Archives.
102. George W. Hunter to Stiles A. Torrance, March 6, 1926, ABC Archives.
103. Stiles A. Torrance to George W. Hunter, April 14, 1926, ABC Archives.
104. Stiles A. Torrance to George W. Hunter, May 6, 1926, ABC Archives.
105. George W. Hunter to Stiles A. Torrance, May 10, 1926, ABC Archives.
106. W. T. H. Howe to George W. Benton, May 3, 1926, ABC Archives. Howe mistakenly dated this April 3.
107. George W. Hunter, *New Civic Biology, Presented in Problems* (New York: American Book Co., 1926), 250.
108. George W. Benton to Louis B. Lee, February 18, 1926, ABC Archives.
109. Ibid., 238.
110. Hunter, *Civic Biology*, 193.
111. Hunter, *New Civic Biology*, 250.
112. Hunter, *New Civic Biology*, 250–51, and *Civic Biology*, 195–96.
113. Hunter, *Civic Biology*, 404, and *New Civic Biology*, 411.

CHAPTER SEVEN

1. Ronald L. Numbers, *Darwinism Comes to America* (Cambridge, MA: Harvard University Press, 1998), 8–10.
2. W. T. H. Howe to W. W. Livengood, March 19, 1924, American Book Company Archives, Syracuse University Library, Syracuse, NY. "Columbia" refers to Teachers College.
3. "The Teacher and the Truth," *Peabody Journal of Education* 2, no. 6 (May 1925): 331–32.
4. Ibid., 333, 329.
5. Ibid., 331.

6. C. H. Thurber to William Jennings Bryan, November 21, 1923, container 38, William Jennings Bryan Papers, Library of Congress, Manuscripts Division, Washington, DC.

7. William Jennings Bryan to C. H. Thurber, December 22, 1923, container 38, Bryan Papers.

8. Benjamin C. Gruenberg, *Elementary Biology* (Boston: Ginn, 1924), n.p. (preceding title page).

9. Benjamin C. Gruenberg, *Biology and Human Life* (Boston: Ginn, 1925), 4–5 (illustrations), 581 (quote).

10. White quoted in "Evolution Not Mentioned in New Dayton Textbook," *Washington Post*, July 19, 1925, 9.

11. James E. Peabody and Arthur E. Hunt, *Biology and Human Welfare* (New York: Macmillan, 1924), 6.

12. Arthur G. Clement, *Living Things: An Elementary Biology* (Syracuse: Iroquois, 1924), 433–34, 455.

13. Minutes of the state textbook commission, October 6, 1925, folder 2, container 44, Governor Austin Peay Papers, Tennessee State Library and Archives, Nashville.

14. William M. Smallwood, Ida L. Reveley, and Guy A. Bailey, *New General Biology* (Boston: Allyn & Bacon, 1929), 646–49.

15. Longfellow quoted in ibid., 646.

16. Ibid., 167–68 (on dinosaurs), 226–30 (on mammals), 699–700 (on eugenics), 706 (quotation).

17. William M. Smallwood, Ida L. Reveley, and Guy A. Bailey, *Practical Biology* (Boston: Allyn & Bacon, 1916), 123.

18. William H. Atwood, *Biology* (Philadelphia: Blakiston's, 1927), 222 (common ancestry), 496 (quotation).

19. Alfred C. Kinsey, *Introduction to Biology* (Philadelphia: J. B. Lippincott, 1926), 196–97.

20. Jeffrey P. Moran, *The Scopes Trial: A Brief History with Documents* (Boston: Bedford/St. Martin's, 2002), 51.

21. Donna J. Drucker to author, August 31, 2010. See also Donna J. Drucker, "Creating the Kinsey Reports: Intellectual and Methodological Influences on Alfred Kinsey's Sex Research, 1919–1953" (Ph.D. diss., Indiana University, 2008).

22. C.E.P. [Carleton E. Preston], review of *New Introduction to Biology*, by A. C. Kinsey, *High School Journal* 17, no. 6 (October 1934): 219.

23. Carleton E. Preston, "The Science Column," *High School Journal* 13, no. 4 (April 1930): 181.

24. J.M.S. [Jesse M. Shaver], review of *New Introduction to Biology*, by A. C. Kinsey, *Peabody Journal of Education* 5, no. 1 (July 1927): 55.

25. *List of Textbooks Approved by State Textbook Commission of Kentucky, 1930–1935* (Frankfort, KY: State Journal Co., 1932), 35.

26. *Textbook Regulations Containing Texas Textbook Law, General Information Regarding the Law and Lists of State Adopted Textbooks* (Austin, TX: State Department of Education, 1934), 43.

27. Drucker, "Creating the Kinsey Reports," 44.

28. Ronald P. Ladoucer to author, July 14, 2010. See also Ronald P. Ladoucer, "Ella Thea Smith and the Lost History of American High School Biology Textbooks," *Journal of the History of Biology* 41, no 3 (September 2008): 435–71.

29. George A. Plimpton, address to firm meeting, January 1926, George A. Plimpton Papers, Rare Book and Manuscript Library, Columbia University, New York.

30. "The Teacher and the Truth," 331.

31. Copies of both C. H. Thurber to William Jennings Bryan, November 21, 1923, and Bryan to Thurber, December 22, 1923, can be found in a folder labeled "C. H. Thurber, 1925" in the

Plimpton Papers. Though no letter accompanies the copies and there is no indication when they were made, their placement within the folder suggests that they were forwarded to Plimpton by Thurber around September 1925.

32. "Biology and Human Welfare by Peabody and Hunt," memorandum enclosed with James E. Peabody to Henry Fairfield Osborn, October 3, 1925, box 1291, Henry Fairfield Osborn Papers, Central Archives, Special Collections, American Museum of Natural History, New York.

33. E.J.A. [E. J. Ashbaugh], "With the Textbook as a Text," *Educational Research Bulletin* 5, no. 16 (November 3, 1926): 338.

CHAPTER EIGHT

1. "Monkey Talk Hits Book Publisher," *Chattanooga Times*, June 5, 1925, 1.

2. Judith V. Grabiner and Peter D. Miller, "Effects of the Scopes Trial: Was It a Victory for Evolutionists?" *Science* 185 (1974): 832.

3. Gerald Skoog, "Topic of Evolution in Secondary School Biology Textbooks, 1900–1977," *Science Education* 63, no. 5 (1979): 628.

4. David E. Moody, "Evolution and the Textbook Structure of Biology," *Science Education* 80, no. 4 (December 1998): 396. See also Dorothy B. Rosenthal, "Evolution in High School Biology Textbooks: 1963–1983," *Science Education* 69, no. 5 (October 1985): 637–48; Gerald Skoog, "The Coverage of Evolution in High School Biology Textbooks Published in the 1980s," *Science Education* 68, no. 2 (April 1985): 117–28.

5. George Gaylord Simpson, "Evolution and Education," *Science* 187, no. 4175 (February 7, 1975): 389–90.

6. Adam R. Shapiro, "Between Training and Popularization: Regulating Science Textbooks in Secondary Education," *Isis* 103, no. 1 (March 2012): 99–110.

7. John Dewey, *Democracy and Education* (New York: Macmillan, 1916), 257.

8. Robert J. Richards, *Darwin and the Emergence of Evolutionary Theories of Mind and Behavior* (Chicago: University of Chicago Press, 1987), 506–8.

9. Simpson, "Evolution and Education," 389–90. See also Ronald P. Ladoucer, "Ella Thea Smith and the Lost History of American High School Biology Textbooks," *Journal of the History of Biology* 41, no 3 (September 2008): 436n.

10. Grabiner and Miller, "Effects of the Scopes Trial," 834. This claim that Kinsey's and Smith's books "either sold poorly or became extinct" is repeated in Randy Moore, "The Lingering Impact of the Scopes Trial on High School Biology Textbooks," *BioScience* 51, no. 9 (2001): 792. For a more complete (and positive) assessment of Kinsey's textbook sales, see Donna J. Drucker, "Creating the Kinsey Reports: Intellectual and Methodological Influences on Alfred Kinsey's Sex Research, 1919–1953" (Ph.D. diss., Indiana University, 2008), 44.

11. Skoog, "Topic of Evolution in Secondary School Biology Textbooks," 622; Grabiner and Miller, "Effects of the Scopes Trial," 836.

12. Ladoucer, "Ella Thea Smith and the Lost History of American High School Biology Textbooks," 435.

13. Arnold B. Grobman, *The Changing Classroom: The Role of the Biological Science Curriculum Study* (Garden City, NY: Doubleday, 1969), 10; John L. Rudolph, *Scientists in the Classroom: The Cold War Reconstruction of American Science Education* (New York: Palgrave Macmillan, 2002), 137–64.

14. Rudolph, *Scientists in the Classroom*, 34–101.

15. Shapiro, "Between Training and Popularization," 102–4.

16. Albert Wertheim, "The McCarthy Era and American Theatre," *Theatre Journal* 34, no. 2 (May 1982): 221–22; Constance Areson Clark, "Evolution for John Doe: Pictures, the Public, and the Scopes Trial Debate," *Journal of American History* 87, no. 4 (March 2001): 1277; Randy

Moore, "Creationism in the United States: VII, The Lingering Impact of 'Inherit the Wind,'" *American Biology Teacher* 61, no. 4 (April 1999): 246; Edward J. Larson, *Summer for the Gods: The Scopes Trial and America's Continuing Debate over Science and Religion* (Cambridge, MA: Harvard University Press, 1997), 238–44.

17. Vassiliki Betty Smocovitis, "The 1959 Darwin Centennial Celebration in America," *Osiris*, 2nd ser., 14 (1999): 274–323. For an example of how discussion of Darwinism increased around 1960, see also the Google Ngram at http://ngrams.googlelabs.com/graph?content=darwinism&year_start=1800&year_end=2000&corpus=0&smoothing=1] (accessed February 3, 2011).

18. George Gaylord Simpson, "One Hundred Years without Darwin Are Enough," *Teachers College Record* 60 (1961): 617–26. Simpson credited the geneticist H. J. Mueller with the phrase that gave the article its title.

19. Ronald L. Numbers, *The Creationists*, expanded ed. (Cambridge, MA: Harvard University Press, 2003), 265.

20. *Epperson v. Arkansas*, 393 U.S. 97 (1968).

21. *Everson v. Board of Education*, 330 U.S. 1 (1947).

22. Numbers, *The Creationists*, 271.

23. For more on the different varieties of creationism, see Numbers, *The Creationists*.

24. *McLean v. Arkansas Board of Education*, 529 F. Supp. 1255 (1982). Philosophers of science have debated whether it is correct to classify creation science as *not* science or as a type of science that happens to be incorrect. For an example of this debate, see Michael Ruse, "Creation Science Is Not Science," in *Philosophy of Science: The Central Issues*, ed. Martin Curd and J. A. Cover (New York: Norton, 1998), 150–60; Larry Lauden, "Commentary on Ruse: Science at the Bar—Causes for Concern," in ibid., 161–66; and Michael Ruse, "Response to Lauden's Commentary: Pro Judice," in ibid., 167–73.

25. *Kitzmiller et al. v. Dover Area School District*, 400 F. Supp. 2d 707 (2005).

26. Percival Davis and Dean H. Kenyon, *Of Pandas and People: The Central Question of Biological Origins* (1989), 2nd ed. (Dallas: Haughton, 1993).

27. *Kitzmiller et al. v. Dover Area School District*, 400 F. Supp. 2d 707 (2005), 32.

28. Ibid.

29. Ibid., 39.

30. Ibid., 40, 41.

31. Shapiro, "Between Training and Popularization," 109.

32. Ronald L. Numbers, *Darwinism Comes to America* (Cambridge, MA: Harvard University Press, 1998), 9–10.

33. *Selman v. Cobb County School Dist.*, 390 F. Supp. 2d 1286 (2005); *Selman v. Cobb County School Dist.*, 449 F.3d 1320 (2006).

34. "Senate Bill 893," http://www.tn.gov/Bills/107/Bill/SB0893.pdf (accessed January 9, 2013). For "monkey bill," see, for example, "Tennessee 'monkey bill' becomes law," *Nature*, April 11, 2012. http://www.nature.com/news/tennessee-monkey-bill-becomes-law-1.10423 (accessed January 9, 2013). Unlike Governor Peay, who issued a statement explaining his reasons for signing the Butler Act, Tennessee Governor Bill Haslam allowed the bill to become law by neither vetoing it nor signing it.

ACKNOWLEDGMENTS

1. I would like to thank the publishers of three of my works for permission to reuse parts of them in revised form in this book: "State Regulation of the Textbook Industry," in *Education and the Culture of Print in Modern America*, edited by Adam R. Nelson and John L. Rudolph, © 2010 by the Board of Regents of the University of Wisconsin System, reprinted courtesy of the University of Wisconsin Press; "The Scopes Trial beyond Science and Religion," by Adam Shapiro, in *Science*

and Religion: New Historical Perspectives, edited by Thomas Dixon, Geoffrey Cantor, and Stephen Pumfrey, © 2010 Cambridge University Press, reprinted with permission; with kind permission from Springer Science + Business Media: *Journal of the History of Biology*, "Civic Biology and the Origin of the School Antievolution Movement," vol. 41, no. 3 (2008): 409–33, by Adam R. Shapiro, © Springer 2008.

Index

Page numbers in boldface refer to illustrations.

academic freedom, 96–99
Adler, Felix, 69–72, 75
Advanced Biology, 57
Agassiz, Louis, 146
alcohol: discussion in textbooks, 75, 77, 119–22, 125; prohibition, 119
Allen, Mrs. C.B., 82–83
Allyn and Bacon, 53, 145–46, 151
American Book Company (ABC): book trust, 16–20, 70, 79; conflict with Ginn, 26–29, 44; formation, 22–29; Mississippi bribery scandals, 14–17, 26, 36–38; organizational structure, 45; publication of *A Civic Biology*, 62–65; Washington bribery scandal, 26
American Civil Liberties Union (ACLU): Henry R. Linville, 70; Scopes trial involvement, 91–92, 96, 98–99; teacher loyalty oaths, 70
American Federation of Teachers, 70, 96
American Teacher, 70, 76
antievolution: laws 7, 12, 37–38, 66, 80–91, 161; religious reasons for, 51–52, 102–4, 107; school antievolution, 65, 86, 88, 107, 114, 150
antievolution laws: Florida, 81; Oklahoma, 81, 90; Tennessee. *See* Butler Act
Are Our Schools in Danger?, 26, 28
A. S. Barnes and Company, 23–24
Ashbaugh, E. J., 154
atheism: Darwinism equated with, 103–4, 138; laws prohibiting atheist teachers, 88–89
Atwood, William H., 56–57, 121, 147
authority: of experts, 84, 128, 151, 155; of science, 49; of textbooks, 41–43, 135, 158, 164–65

bachelors, 29
Bailey, Guy H., 53, 56–57, 66, 78, 121, 145, 149, 151, 155
Baltimore Sun, 9
Barnes, John C., 23
Benton, George W., 44–45, 58–60, 63, 111–12, 115–32, 135, 151
Bible Crusaders, 125
Bible Wars, 32
Bigelow, Maurice A., 56–57
Bilbo, Theodore, 14–16, 35, 37
Biological Field Club, DeWitt Clinton High School, 68–**69**
Biological Sciences Curriculum Study (BSCS), 160–61
biology (school subject): replaces botany and zoology 61, 64–68, 74–78, 118; shift to tenth grade, 78, 116; urban-rural split, 36, 54, 66, 75, 78–81, 114–15, 149
Biology and Human Life, 85, 119, 125–26, 139–40, 143, 146–47, 149, 151–52, 155
Biology and Human Welfare, 85, 119, 140, 143, 147, 149, 152, 154
Biology for High Schools, 78
Boston, 1915 textbook adoption, 62–65, 76, 86, 114
Bryan, William Jennings: belief in antievolution, 103–5; correspondence with Charles Thurber, 61, 138–39, 151, 153, 158; examination by

Bryan, William (*cont.*)
Clarence Darrow, 105–6; on literalism, 103; Presbyterianism, 87; speech at Scopes trial, 101, **102**, 103
Butler, John Washington, 81–82, 88–91
Butler Act, 66, 88–90; interpretation by Scopes defense, 95–97; passage, 66, 80–84; repeal, 162; signing statement by Peay, 7, 38, 40, 81, 91, 96

Caldwell, Otis, 56, 70–75, 78, 165
California: communication lag with New York, 111–12; constitutional convention 29–34; teaches evolution, 154, textbook printing, 29–34, 37
Carnegie, Andrew, 51
Chattanooga Times, 92
chemistry, as argument against evolution, 87
Chicago Tribune, 26, 111
Chicago, University of. *See* University of Chicago
Civic Biology: adoption in Boston, 62–65; on evolution, 108–9; quoted by Bryan, 101–**2**; read by John Scopes, 107–8
Civic Science in the Home and Community, 116, 130
Clement, Arthur G.: controversy over 1924 publication of *Living Things*, 53–58, 85; post-Scopes revision of *Living Things*, 141, **142**, **143**, 148–51, 155
Columbus Journal, 111
Committee for the Reorganization of Science Education, 70–71
common descent of animals, 76–77
conflict of science and religion: as framed by historians, 4–7, 12–13, 132, 138, 165; as invoked in Scopes trial, 86–89, 92, 97–107
copyright laws, 21–22, 25
creationism, 107, 162–63
creation science, 162–63
Cumberland Valley, 81, 88

D. Appleton and Company, 24
Darrow, Clarence: on Biblical literalism, 101; exchange with William Jennings Bryan at Scopes trial, 95
Darwin, Charles, biography in textbooks; 141–46, **142**, **143**
Dayton, Tennessee: quintessentially American, 6–13, **9**; school, 3, 8, 66, 91, 95; seeking publicity, 91–93
Democracy and Education, 73
DeWitt Clinton High School, New York: 57, 67–70, 74; Biological Field Club, 68–**69**
Dewey, John: educational theories, 73–74; New York City Teachers Union, 70
Dillman, Louis B., 129
Domer, David S., 6
Donovan, Herman L., 41
Dover, Pennsylvania. See *Kitzmiller et al. v. Dover Area School District*
Draper, John, 106
DuBois, W.E.B., 10
due process, 96–97, 161

Eastern Kentucky State Teachers College, 41
Education: Intellectual, Moral, Physical, 72
Edwards Hotel, Jackson, Mississippi, 14, 36
Elementary Biology, 117, 139–40
Elements of Biology, 74
Epperson v. Arkansas, 161–62
Essentials of Biology, 74, 78
Establishment Clause, 97
Ethical Culture Society, 68–71
eugenics: in *A Civic Biology*, 62–63, 75, 77; education as, 72; in other textbooks, 139, 141, 146; reason for antievolution, 51, 64
even exchange, 19, 26–27
evolution: in pedagogy, 72–74, 165; religious reactions, 65, 81, 86, 101–4, 162–63; use of word in biology textbooks, 137–48, 151–56, 158–60
evolutionary tree, diagram in textbooks, 108–10, **109**, 128, 130, 140, 150
experts: testimony in Scopes trial, 84–86, 97, 100–101, 105; in textbook adoptions 35–36, 40–42, 46–47, 64, 71

Fairchild, Jarvis R., 63, 115, 126
Ferguson, Miriam "Ma," 148
Fitzpatrick, Elizabeth T., 57
Flood, as Biblical rationale for young earth, 162
Foe to American Schools, A, 27
freedom of speech: Argument in Scopes trial, 96; academic freedom in schools, 98
freedom of religion, Argument in Scopes trial, 97; Tennessee constitution, 97
Fundamentals, The, 65

Gates, George A., 27–28
General Biology (Sedgwick and Wilson), 67–68
General Education Bill, 80–83
general science, 61, 70–74, 78–79, 116
Genesis, cited as basis for antievolutionism, 88–89, 99, 107, 137, 162–64
Ginn, Edwin: early career, 22, 47, 52; founder of Ginn and Company, 19, 22, 28; philanthropy, 50–52
Ginn and Company: after Edwin Ginn, 28, 49; Gruenberg's Scopes trial testimony, 84–85; publisher of biology textbooks, 54–57, 85, 92, 117, 122,

139–40; rivalry with ABC, 18–20, 24–28, 44; trade organizations, 23–24
glossary, evolution mentioned in, 147, 155
Grinnell College, 27
Gruenberg, Benjamin Charles: asked to testify at Scopes trial, 84–86; curriculum reform, 69–70; response to Clement controversy, 54–58; sex education, 148; textbook author, 54, 57, 68, 85, 92, 108, 117–20, 126
Gruenberg, Sidonie Matsner, 68

Hall, G. Stanley, 73–74, 160
Harned, Perry L., 2–3,19, 38, 81–85
Harper, Fletcher Jr., 25
Harper and Brothers, 25
Hays, Arthur G., 95, 99, 105
Hazen, Annah P., 122
Hell's Kitchen, 67
homologous structures, diagram in textbooks, **133**, 146
horse, evolution diagram in textbooks, 123, 128, 130, 133
Howe, W.T.H. (William Thomas Hildrup), 14–15, 35–40, 44–45, 58–59, 61, 121, 127–31, 137
human descent from nonhumans, 108, 110, 120, 133
Hunt, Arthur E., 41, 56–57, 85, 96, 119–20, 140, 143, 149, 150, 158
Hunter, George William: author of *Civic Biology*, 74–77; Biological Field Club founder, 68–**69**; creation of *New Civic Biology* 111–34; early life, 68; Knox College professor, 78, 115, 117, 127; relationship with American Book Company, 60, 124–27; University of Chicago student, 68
Hyde Park High School, 123

indexes, use of evolution in, 120, 122, 141, **144**, **145**, 147–48
Inherit the Wind, 161
In His Image, 104
intelligent design (ID), 12, 163–64
Introduction to Biology, 147–50
Iroquois Publishing Company: founding, 52–54; revision of Clement's *Living Things*, 141–47, 151
Ivision, Blakeman and Company, 27

J. B. Lippincott, 148
Johns Hopkins University, 100
Jones, John E., III, 12, 163
Julia Richmond High School for Girls, Brooklyn, 54

Kingdom Publishing Company, 27–28
Kingsley, Clarence Darwin, 69–71, 75
Kingston New Leader, 111
Kinsey, Alfred E., 147–50, 155, 160
Kitzmiller et al. v. Dover Area School District, 12, 163, 165
Kline, Nelson S., 57
Knox College, 78, 115, 117, 127

Lee, Louis B., 59, 118, 121–31
Lewis, Sinclair, 8
Linville, Henry R: 55–57, 67, 70–76, 98
Lippincott. *See* J. B. Lippincott
literalism: basis for antievolution, 98, 162, 164–65; caricature by evolutionists, 92–93, 100–107, 137–38; inspiration versus interpretation, 103; in reading of textbooks, 110, 156, 159
Littlejohn, Joseph, 14, 37
Livengood, W.W. (William Winfred), 44, 59–60, 63, 117–131, 135
Living Things: An Elementary Biology: 1924 publication controversy; 1925 revision 141, **142–45**
Longfellow, Henry Wadsworth, 146
Los Angeles Herald, 111
Louis, Henry L., 14–16, 37–38, 44
loyalty oaths, 70, 98

Macmillan, 36, 56, 85, 119, 140, 147, 154
Main Street, U.S.A., 8–11
majoritarianism, 99
Malone, Dudley Field, 95–96, 99–100; rejoinder to Bryan, 104
Mandl, Max, 57
Mann, Horace, 20–22
Matthews, Shailer, 97
McFarlane, Charles, T., 58
McGuffey Readers, 24, 30
McLean v. Arkansas Board of Education, 162–63
Mencken, H.L., 9–10
Mendel, Gregor, 128
Metcalfe, Maynard, 100
Minneapolis Morning Tribune, 111
Mississippi: state teachers conference of 1923, 39, 44, 61; textbook printing scandal, 14–15, 36
Modern Biology, 149
modernism, 94
Modern Times and the Living Past, 39
Mohonk Conferences on International Arbitration, 51
monkeys, misunderstood human ancestry from, 119–20, 138
Moon, Truman J., 41, 56–57, 121, 149
Moral Instruction of Children, The, 69, 72–73

Neal, John R., 85
Nebraska, 6

New Biology, 145
New Civic Biology, 36, 111–36, 140–41, 147, 149–50, 153–55, 158–59
New Essentials of Biology, 5, 78–79, 112–16, 123
New General Biology, 145–46, 149, 155
New Introduction to Biology, 150
New York City Teachers Union, 70–71
New York State Department of Education, 54–55, 119
New York Times, 83, 87–88, 93
Nihart, W.F., 14–15, 36

Oberlin College, 100
objectivity, 42
Of Pandas and People, 163–64
"One Hundred Years without Darwin Are Enough," 161
On the Origin of Species, 5, 12, 24, 86, 140, 142, 161
Osborn, Henry Fairfield, 146, 154
O'Sullivan, James, 29–30, 33–34

pamphlets: at the Scopes trial, 8, 94, 96; use by textbook publishers, 27–28, 34, 44, 50, 154
paper costs, 1, 5
Peabody, James E., 56–57, 70–71, 85, 96, 119–20, 140, 143, 150, 154, 158
Peabody College, Nashville, Tennessee, 35–36, 41, 127–28, 136–38
Peabody Journal of Education, 128, 137–38, 149, 151
Peay, Austin M.: Butler Act signing statement, 7, 37–40, 89, 96, 107; general education bill, 80–84; postponement of 1924 textbook adoption, 1–3
Pennsylvania intelligent design trial. See *Kitzmiller et al. v. Dover Area School District*
Pinkerton Detective Agency, 14, 37
Plimpton, George A., 23, 28, 46–47, 151
Presbyterian Church in the U.S.A., 87
Preston, Carleton, E., 149
primitive human ancestors, 77, 108–10, 133
printing of textbooks by states, 15–16, 29–34
Prohibition (of alcohol), 119
Publishers' Alliance, 24
Publishers Board of Trade, 23
Publishers' Weekly, 24–25, 30
pussy-footing, 119–20

racism, and antievolutionism, 10
recapitulation, 72–74
Regents Examination, 53
Reveley, Ida L., 53, 56–57, 66, 78, 121, 145, 149, 155
Rhoads, McHenry, 123, 131, 136, 151
Riley, William Bell, 98
Robb, Mary E., 123–24, 129
Robinson, Frank, 96, 108
rural school problem, 80

salesmanship, 20–21, 48–52
School Book Publishers Association, 27
science and religion: conflict thesis, 87–89; demarcation 88, 98; interpretation in Scopes trial, 97–98; trope of, 88
Scopes, John T.: innocence, 107–10; memoir, 91; on reading *A Civic Biology*, 7, 108
Sedgwick, William T., 67–68
See and Say Series, 52
sex education, 64–65, 77, 148
Shaver, Jesse M., 36, 127–32, 136, 149, 151, 153
Simpson, George Gaylord, 158, 161
Smallwood, William M.,
Smith, Ella Thea, 149, 160
social evolution, 49, 51, 74
Southworth, Edward F., 51–53, 141–43, 151
Spencer, Herbert, 51, 72, 160
Sputnik, 160
statewide adoption, 18, 31, 32, 33, 35
stickers in textbooks, **136**, 164–65
St. John, Charles J., 39, 42, 44, 61
St. Louis Star, 111
suicide, as undesirable, 25, 118
syphilis, 62–63

Taft, Robert A., 15
Teachers College, Columbia University, 35–36, 41, 57–58, 68–70, 73, 137
Texas: ABC exclusion from, 28; antievolutionism in, 148
theistic evolution: Bryan's opposition to, 103–4, 164; as part of Scopes trial, 90, 97–98, 162
Thorndike, Edward L., 73
Thurber, Charles, 61, 138–40, 150–53, 158
Torrance, Stiles A., 126–31, 135
Trafton, Gilbert H., 56

uniform adoption. *See* statewide adoption
University of Chicago, 35, 64, 68, 97
University of Tennessee, 82
urban-rural differences, 10–11, 36, 54, 75, 78–81, 114–15, 149–50

vacations, 118, 120–22, 132
vagueness, 96–97
Van Antwerp, Bragg and Company, 27

Waggoner, Harry D., 149
WGN radio, 93, **102**
Wheat, Frank M., 57
White, Andrew Dickson, 106–7, 138
White, Walter, 95, 140
Whitman, Charles O.: Clark University, 74; University of Chicago, 68; Woods Hole, 68
Whitman, Walter G., 62, 70, 74–75, 78, 116, 130
Who Is Edwin Ginn?, 27
Why Dayton of All Places?, 8–**9**
Wilson, Edmund B., 67–68
Woods Hole Biological Station, 68
World Christian Fundamentals Association, 98
World Peace Foundation, 51

young earth creationism, 107, 162